Reviews of

107 Physiology, Biochemistry and Pharmacology

Editors

P. F. Baker, London · H. Grunicke, Innsbruck
E. Habermann, Gießen · R. J. Linden, Leeds
P. A. Miescher, Genève · H. Neurath, Seattle
S. Numa, Kyoto · D. Pette, Konstanz
B. Sakmann, Göttingen · W. Singer, Frankfurt/M
U. Trendelenburg, Würzburg · K. J. Ullrich, Frankfurt/M

With 23 Figures and 6 Tables

Springer-Verlag
Berlin Heidelberg GmbH

ISBN 978-3-662-31071-7   ISBN 978-3-540-47715-0 (eBook)
DOI 10.1007/978-3-540-47715-0

Library of Congress-Catalog-Card Number 74-3674

© by Springer-Verlag Berlin Heidelberg 1987
Originally published by Springer-Verlag Berlin Heidelberg New York in 1987
Softcover reprint of the hardcover 1st edition 1987

2127/3130-543210

Contents

Indexed in Current Contents

Rev. Physiol. Biochem. Pharmacol., Vol. 107
© by Springer-Verlag 1987

Renal Prostaglandins and Leukotrienes

LARRY A. WALKER[1] and JÜRGEN C. FRÖLICH[2]

Contents

---

[1] Research Institute of Pharmaceutical Sciences, University of Mississippi, Mississippi 38677, USA
[2] Department of Clinical Pharmacology, Medical School, D-3000 Hannover, FRG

1 Renal Prostaglandin Synthesis

1.1 Introduction

The quantitatively most important precursor of prostaglandins (PGs), thromboxanes (TXs), prostacyclin (PGI_2), and hydroxylated fatty acids in the kindey is arachidonic acid (5,8,11,14 eicosatetraenoic acid). Arachidonic acid is converted by the enzyme complex cyclooxygenase into cyclized compounds (prostanoids), or by lipoxygenase and mixed-function oxidase into hydroxylated fatty acids with three or four double bonds, including the leukotrienes (LTs) (Fig. 1). Other C_{20} fatty acids can serve as precursors, such as 8,11,14 eicosatrienoic or 5,8,11,14,17 eicosapentaenoic acid. Eicosatrienoic acid is present in mammalian tissues in much smaller quantities than arachidonic acid, and its metabolites (PGE_1, $PGF_{1\alpha}$, TXB_1) are very much less abundant than the arachidonic acid metabolites (Knapp et al. 1978). Although it was recently shown that LTs can be formed from all-cis-5,8,11,14,17 eicosapentaenoic acid (Hammerström 1980), this fatty acid is a poor substrate for cyclooxygenase (and probably for lipoxygenase as well), and, if converted to biologically active oxygenated metabolites the amounts formed are exceedingly small (Sies et al. 1984; Fischer and Weber 1984). Also, the renal C_{22} precursor fatty acid 7,10,13,16 docosatetraenoic acid has been poorly studied but appears not to contribute significantly to renal synthesis of oxygenated fatty acids (Tobias et al. 1975). Thus, arachidonic acid remains the most abundant and best-studied precursor and will alone be considered in this review. It gives rise to the bisenoic prostanoids, i.e., prostanoids with two double bonds, and to LTs (see Fig. 1).

Because of the recent discovery of the lipoxygenase pathway in the kindey (Winokur and Morrison 1981; Oliw and Oates 1981; Jim et al. 1982), it is suggested that the term "eicosanoid" (Corey et al. 1980) rather than "prostanoid", which only refers to cyclized products of arachidonic acid, is the most appropriate to use when discussing arachidonic acid metabolism in the kidney, as it encompasses both cyclooxygenase and lipoxygenase products.

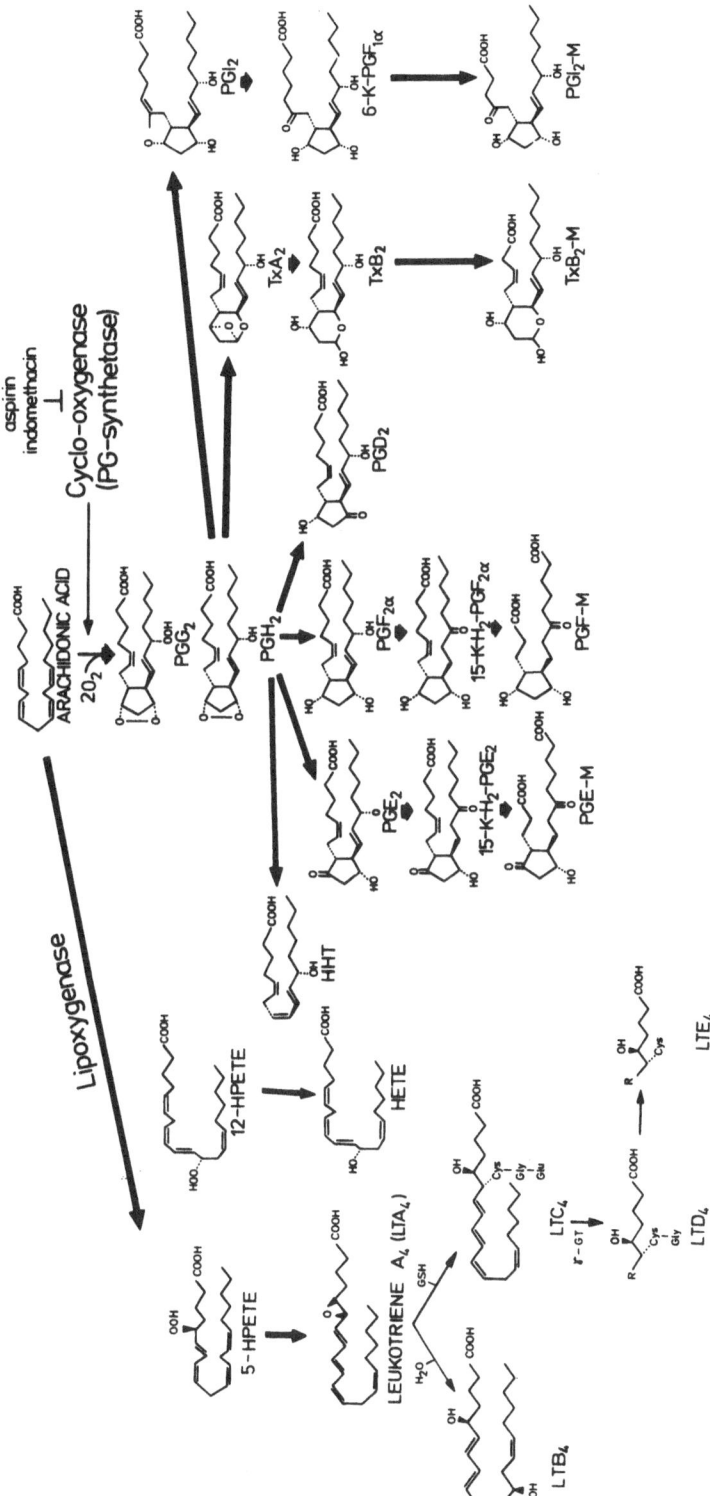

Fig. 1. Major biochemical pathways of eicosanoid synthesis and metabolism

1.2 Historical Background

Biologically active material in extracts of human prostate tissue was described at the beginning of this century (Battez and Goulet 1913; Kurzrok and Lieb 1930). This material was identified as different from autacoids then known (von Euler 1934; Goldblatt 1935) and named prostaglandin by U.S. von Euler (1935). The interesting pharmacologic effects on smooth muscle (gut, uterus) in vitro and on blood pressure (von Euler 1934) were not fully investigated at that time, largely owing to a lack of active material.

Initial attempts at elucidation of the structure of the biologically active material isolated from sheep prostate gland were made by Bergström and Sjövall (1960). Elemental and mass spectrometric analysis established that the two compounds isolated were unsaturated acids with molecular weights of 356 (PGF_1) and 354 (PGE_1), with elemental compositions $C_{20} H_{36} O_5$ (PGF_1) and $C_{20} H_{34} O_5$ (PGE_1). Further work by Bergström et al. 1962c; 1963b) produced mass spectra of PGE_1, $PGF_{2\alpha}$, and PGF_{1a} and the correct chemical structures were proposed (Bergström et al. 1962). Two additional PGs, PGD_2 and PGE_3, were identified by similar methods (Bergström et al. 1962b, Samuelsson (1963). Once the exact chemical nature of the PGs had been established, it became much easier to show whether or not compounds with similar biological properties were PGs. In a study of various tissues from sheep, it was discovered that the lung contained a smooth muscle-stimulating substance (Bergström et al. 1962a). Extraction from 440 kg of swine lung yielded a few milligrams of this compound, quite enough to give a complete mass spectrum which identified it unequivocally as PGE_2. This was the first reported occurrence of PGs outside the reproductive system and showed in an exemplary way the approach to definitive identification of suspected PG compounds. Gas chromatography-mass spectrometry was utilized for this work as well as for the demonstration of $PGF_{3\alpha}$ in bovine lung (Samuelsson 1964b) and brain (Samuelsson 1964b), and of PGs in human seminal plasma (Hamberg 1966), including the novel 19-hydroxyprostaglandins. After the exact chemical structure of the PGs had been established, it became possible to address the questions of their precursor, the mechanism of biosynthesis, metabolism, and chemical synthesis.

In 1963 Bergström et al. (1963a) and in 1964 van Dorp et al. discovered that arachidonic acid (5,8,11,14 eicosatetraenoic acid) was the precursor of the bisenoic PGs PGE_2 and $PGF_{2\alpha}$; all-cis 8,11,14 eicosatrienoic acid the precursor of PGE_1 and $PGF_{1\alpha}$ (Samuelsson 1965; Nugteren and Hazelhof 1973); and all-cis 5,8,11,14,17 eicosapentaenoic acid the precursor of PGE_3 and $PGF_{3\alpha}$ (Nugteren and Hazelhof 1973). This discovery led to the postulation of intermediates in PG synthesis (Samuelsson 1965), which were subsequently isolated (Nugteren and Hazelhof 1973; Hamberg and Samuelsson 1973). It was soon recognized that these intermediates, called cyclic endoperoxides or PGG_2 and PGH_2, had powerful biological activities independent of their conversion to PGE_2 and $PGF_{2\alpha}$.

A significant advance in our understanding of the PG system was the discovery of the inhibitory effect of aspirin on PG synthesis (Vane 1971; Smith and Willis 1971). This effect, shared by a large number of other nonsteroidal anti-inflammatory drugs, provided a tool for the study of the effects of PGs synthesized endogenously and thus helped to elucidate their roles in organ function. This tool proved to be of great help, the more so because antagonists to PGE_2 or $PGF_{2\alpha}$ were not available.

Later, a new unstable product formed from the cyclic endoperoxide PGG_2 by blood platelets was identified, together with is more stable, biologically inactive product (Hamberg et al. 1974). Because of its prominent platelet-aggregating property, the active compound was named thromboxane A_2 (TXA_2) (Hamberg et al. 1975).

In a systematic study of biconversion of PGG_2 by microsomes of various organs, it was discovered that incubation with arterial microsomes gave rise to a labile compound which inhibited platelet aggregation and reduced the tone of blood vessels (Moncada et al. 1976a). This compound was initially labeled PGX. Because PGX could also be generated from rat stomach homogenates (Pace-Asciak et al. 1976), it was thought to be related to a cyclic ether (Pace-Asciak and Wolfe 1971) formed from arachidonic acid (Pace-Asciak 1976b) or cyclic endoperoxides (Pace-Asciak 1976a) in that same tissue. The compound was identified, synthesized (Johnson et al. 1976), and named prostacyclin (PGI_2) because of its cyclic structure. It has received considerable attention because it is a vasodilator which prevents platelet aggregation and is the main arachidonate metabolite of the vascular endothelium (Bunting et al. 1976; Moncada et al. 1976b).

The metabolism of PGs, especially that of PGE_2 and $PGF_{2\alpha}$, has been studied in vivo and in vitro. PGs are rapidly converted by most tissues to their 15-keto-13,14-dihydro analog by 15-hydroxy-PG-dehydrogenase and 13-reductase (Anngard and Larsson 1971). Further catabolism of PGs includes interconversion of hydroxyl and keto groups at C-9 (Pace-Asciak 1975; Lee and Levine 1974), and omega or beta oxidation (Hamberg and Samuelsson 1971). Metabolism of PGI_2 has been studied in detail in animals and man, and its major circulating metabolites (Rosenkranz et al. 1981b) and urinary metabolites (Rosen-kranz et al. 1981a; Brash et al. 1983) have been identified. Much of the metabolism appears to occur in the tissue of origin, while the lung acts as a filter and removes any PGs that might escape local metabolism (Ferreira and Vane 1967). Metabolism occurs at a very rapid rate and the metabolites are usually biologically inactive.

1.3   General Properties of the Eicosanoid System

PGs are not stored intracellularly. They are generated rapidly after an appropriate stimulus or excision of the tissue, exert their effects, and are quickly metabolized. For these reasons, PGs have been termed local or tissue hormones or autacoids, and, indeed, no circulating PG in concentrations likely to cause biological effects has as yet been found under physiologic conditions.

Early reports on the synthesis of PGA_2 in the renal medulla (Lee et al. 1965) led to the suggestion that PGA_2 might function as a cirulat-ing hormone because, in contrast to PGE_2, it was found to cross the pulmonary vascular bed largely unmetabolized (McGiff et al. 1969). Furthermore on the basis of radioimmunoassay, PGA_2 was described as occurring in the circulation in concentrations of several nanograms per milliliter (Zusman et al. 1973; Zusman et al. 1974), an amount suf-ficient to cause biological effects. Because PGE_2 can be dehydrated to form PGA_2 in vitro under the acidic conditions commonly employed to extract PGs from biological sources, we wondered what amount of the PGA_2 observed in renal medulla might actually have arisen by this route. We found that all the PGA_2 detected in renal medullary homo-genates of rabbit and human kidney originated from the dehydration of PGE_2 during the extraction process (Frölich et al. 1975a). In our subsequent work on levels of PGA_2 in human circulation, utilizing gas chromatography-mass spectrometry, we found the PGA_2 levels were indistinguishable from zero (Frölich et al. 1975b).

PGE_2 levels in the circulation have been detected by radioimmuno-assay in amounts that could have biological importance. However, on the basis of measurements of the major urinary and circulating metabolites, it was shown that the true levels are much smaller and in the range of a few picograms per milliliter (Hamberg and Samuelsson 1971, 1973). Under conditions of shock, however, PGE_2 levels were found by bio-assay to be elevated to levels of several nanograms per milliliter (Jakschik et al. 1974), and this was subsequently confirmed by gas chromato-graphy-mass spectrometry (Frölich 1977).

PGI$_2$ has also been described as a PG appearing in the arterial circulation after release from the lung (Gryglewski 1979). We have studied the metabolism of PGI$_2$ in man after its intravenous infusion and found that dinor-6-keto-PGF$_{1\alpha}$, dinor-13,14,-dihydro-6,15-diketo-20-carboxyl-PGF$_{1\alpha}$, and 6-keto-PGF$_{1\alpha}$ are its major urinary metabolites (Rosenkranz et al. 1980). Measurement of urinary 2, 3-dinor-6-keto-PGF$_{1\alpha}$, under basal conditions reveals levels of approximately 400 ng/24 h in normal volunteers (B. Rosenkranz, unpublished observations). Because urinary 2, 3-dinor-6-keto-PGF$_{1\alpha}$ represents about 7% of infused PGI$_2$, one can calculate that total body PGI$_2$ synthesis is about 6 μg/24 h, corresponding to about 0.06 ng/kg/min. No biological effects have as yet been reported when PGI$_2$ is infused at that rate. Fischer et al. (1982) also reported very low levels of 6-keto-PGF$_{1\alpha}$ in normal adults. Thus, at present, a biologically significant effect on renal function cannot be attributed to circulating PGI$_2$ or other PGs. Renal PG effects are likely due to local synthesis under basal conditions.

A similar approach to measurement of PG metabolites has recently been taken by estimating the renal clearances of ^{3}H-PGE$_2$, ^{3}H-6-keto-PGF$_{1\alpha}$, and ^{3}H-TXB$_2$ infused into the renal artery or brachial vein in man (Smith et al. 1982). These studies confirmed that 6-keto-PGF$_{1\alpha}$ given intravenously results in the urinary excretion of 6-keto-PGF$_{1\alpha}$ (Rosenkranz et al. 1980). On the basis of the percentage of urinary excretion of unchanged radioactive PGs and the amount excreted from endogenous synthesis, maximum plasma levels were calculated (Zipser and Martin 1982) and found to be similar to those estimated on the basis of metabolite measurements for PGE$_2$ and PGF$_{2\alpha}$ (Frölich et al. 1979) or direct measurement (Smith et al. 1982).

The release of PGs is not the consequence of discharge from a pool, because the amount released is always much greater than the amount that can be detected in the tissue. This has been shown to be true for the spleen (Gilmore et al. 1968), lung, adrenal, intestine (Ramwell and Shaw 1970), and kidneys (Anggard et al. 1972), and no exception has as yet been reported. The rate of PG formation is quite high. For example, in the renal medulla, a 3000-fold increase in PGE$_2$ has been observed within minutes of excision of the tissue (Anggard et al. 1972).

1.4 Precursor Release

Eicosanoid synthesis requires C$_{20}$ unsaturated fatty acids as substrate. These fatty acids must be present in the nonesterified form before PG synthesis can take place (Lands and Samuelsson 1968; Vonkeman and van Dorp 1968). The intracellular levels of nonesterified fatty acids are extremely low (Kunze and Vogt 1971; Haye et al. 1973; Samuelsson 1972). Initiation of PG synthesis, therefore, depends on the cleavage of the fatty acid from an esterified precursor. The source of the free acid could be intracellular lipid pools such as cholesterol esters, phospholipids, or mono-, di-, and triglycerides. Thus, several enzymes are potentially capable of mobilizing the necessary fatty acid substrate. The suggestion that has received most experimental support is that the nonesterified fatty acid is released from the phospholipid fraction under the influence of a phospholipase (Kunze and Vogt 1971). Indeed, administration of phospholipase resulted in release of PGs (Vogt et al. 1970; Palmer et al. 1973). Perfusion of organs with arachidonic acid results in synthesis of PGs (Kunze and Vogt 1971; Bartels et al. 1970; Palmer et al. 1973; Vargaftig and Dao Hai 1972; Chang et al. 1975) which can be blocked pharmacologically, thus indicating that the enzymatic conversion to PGs proceeds as soon as substrate is available. In the guinea pig lung, infusion of arachidonic acid results in the appearance of PGs in the effluent (Vargaftig and Dao Hai 1972). Bradykinin will also release PGs in this system. Mepacrine, an inhibitor of phospholipase A_2, has no effect on PG synthesis during arachidonic acid infusion but will prevent PG release in response to bradykinin (Vargaftig and Dao Hai 1972).

A more detailed analysis of the release of arachidonic acid from the phospholipids was attempted by labeling platelet lipids with ^{14}C-arachidonic acid (Bills et al. 1976). Ninety-five percent of the radioactivity taken up by the platelets appeared in the platelet phospholipid fraction. More than half of the labeled arachidonic acid was in the phosphatidylcholine fraction. Interestingly, on stimulation of platelet aggregation by thrombin, there was a burst of release of radioactive compounds (mainly TXB_2, HHT, and HETE; see Fig. 1) and a simultaneous proportional decrease in the radioactivity of phosphatidylcholine and phosphatidylinositol, thus indicating that the released arachidonic acid is quantitatively converted by cyclooxygenase and lipoxygenase to oxygenated products. In this study, phosphatidylcholine accounted for the majority of the released arachidonate. Other investigators have suggested phosphatidylinositol as the predominate source (Haye et al. 1973). It appears that the phospholipid source may vary from one tissue to another and may depend on the stimulus for PG synthesis. Also, there is some evidence that the particular phospholipid pool labeled by exogenous arachidonate may vary with the incubation times (Daniel et al. 1981; Schremmer et al. 1979). This point requires further study.

Several schemes for arachidonate release have been suggested. It has been widely accepted that phospholipase A_2 is activated by hormonal or other stimuli in a calcium-dependent fashion to cleave arachidonate from the 2 position of phosphatidylcholine, phosphatidyl ethanolamine, or other phospholipids. Support for this study proposal has come from a variety of studies (see above). Another pathway leading to arachidonate release involves the activation of phosphatidylinositol-specific phospholipase C, followed by the deacylation of the 1,2-diacylglycerol by a diglyceride lipase which cleaves arachidonate from the 2 position and leaves a monoglyceride. This sequence of events occurs in platelets (Bills et al. 1977; Bell et al. 1979) and also in neutrophils (Bell et al. 1979). More recently, it has been proposed (Billah et al. 1980) that in platelets, following the action of phospholipase C, the diacylglycerol is phosphorylated to form phosphatidic acid, which mobilizes calcium (Tyson et al. 1976) and activates phospholipase A_2. This results in the release of arachidonate from phosphatidic acid, phosphatidylcholine, or phosphatidylethanolamine (phosphatidylinositol is generally not a good substrate for phospholipase A_2). It is of particular interest for this review that support for this latter scheme has been obtained in rat renal medullary slices, where angiotensin II was used to stimulate (Benabe et al. 1982). It is to be emphasized that the extension of these results to other tissues or even cell types in a given tissue is premature. Further, the various stimuli used to increase PG synthesis may activate different lipases which act on different substrates. For example, a phosphatidylinositol-specific phospholipase A_2 was found in transformed mouse BALV/3T3 cells (Hong and Deykin 1981).

A number of stimuli for enhanced renal PG synthesis have already been mentioned (angiotensin II, bradykinin); other factors will be mentioned below in the discussion of the various system functions. They include antidiuretic hormone (ADH), ischemia, endotoxin administration, and ureteral obstruction.

1.5 Prostaglandin Cyclooxygenase and Lipoxygenase

The nonesterified arachidonic acid can be acted upon by either of two enzymes, cyclooxygenase (yielding cyclized oxygenated products, the endoperoxides PGG_2 and PGH_2) or lipoxygenase (yielding noncyclized oxygenated products such as HETE and LTs; Fig. 1). The cyclic endoperoxides can be further converted to PGE_2, PGD_2, and possibly $PGF_{2\alpha}$ by isomerases or spontaneous breakdown, or enzymatically to TX A_2 or PGI_2. Thus, cyclooxygenase assumes a central role in the generation of all these compounds. Both lipoxygenase and cyclooxygenase are inhibited by indomethacin, aspirin, meclofenamate, and other nonsteroidal anti-inflammatory agents (Smith and Willis 1971; Sircar et al. 1983; for review see Flower 1974) (Fig. 1).

The mechanism of oxygen incorporation by cyclooxygenase has been studied in detail. It was found that the oxygen of the hydroxyls at C-11 and C-15 of PGE_2 originated from molecular oxygen (Nugteren and van Dorp 1965; Ryhage and Samuelsson 1965); however, the origin of the oxygen function at C-9 remained obscure. In elegant experiments, using a mixture of $^{18}O_2$ and $^{16}O_2$ in the biosynthesis of PGE_1, Samuelsson (1965) demonstrated by mass spectrometric analysis that the oxygen functions at C-9 and C-11 were either both ^{18}O or ^{16}O, thus proving that they must have been derived from the same molecule. Furthermore, he postulated that an endoperoxide intermediate must be formed. This endoperoxide, PGG_2, was subsequently isolated (Nugteren and Hazelhof 1973; Hamberg and Samuelsson 1973) and shown to be converted enzymatically to PGH_2 (Miyamoto et al. 1976).

Cyclooxygenase has been purified from bovine vesicular gland microsomes (Miyamoto et al. 1976; Hember et al. 1975), and it has been shown that cyclization at C-8 and C-12 occurs simultaneously with incorporation of two molecules of O_2 at C-11 and C-15 (Miyamoto et al. 1976). The next step – conversion of the 15-hydroperoxy group by glutathione peroxidase – had been suggested (Nugteren and Hazelhof 1973); enzymatic conversion of the 15-hydroperoxy group was observed even in the absence of this enzyme, and glutathione accelerated this reaction in the absence of the enzyme (Miyamoto et al. 1976).

One outstanding property of cyclooxygenase has been puzzling and annoying, namely, its ability to self-destruct (Smith and Lands 1972). Recent evidence indicates that the cause of this destruction is the presence of radicals formed as the consequence of reductive breakdown of the 15-hydroperoxide which could oxidize the enzyme (Egan et al. 1976). Scavenging of the radicals by phenol, a recognized "stimulator" of cyclooxygenase activity, prevents inactivation of the enzyme (Egan et al. 1976). Whether this enzyme oxidation occurs in vivo and whether there exists a natural protective reducing agent in vivo is unknown.

1.6 PG Isomerases

PGH_2 can be converted into PGE_2, PGD_2, PGI_2, and TXA_2 by isomerases. Each tissue and possibly each cell type has its unique mixture of isomerases. Thus, endothelial cells synthesize predominately PGI_2 and blood-platelets TXA_2 upon stimulation. It is important to realize that at least PGE_2 can also result from the nonenzymatic breakdown of PGH_2 in aqueous solution (Nugteren and Hazelhof 1973; Hamberg and Samuelsson 1973; Hamberg and Fredholm 1976). Furthermore, in in vitro studies utilizing isolated cells or tissue slices, the presence of cofactors can significantly alter the ratio of the various PGs formed. This makes it difficult to interpret in vitro data and to relate them to in vivo biosynthesis of PGs.

Conversion of one PG to another can occur in vivo. Thus, PGD_2 was shown to be converted to $PGF_{2\alpha}$ (Ellis et al. 1979). Also, an enzyme that converts PGE_2 to $PGF_{2\alpha}$ was found in the kidney (Hamberg and Israelsson 1970; Lee and Levine 1974). In vivo changing ratios of PGE_2 to $PGF_{2\alpha}$ have been attributed to changes in the activity of the enzyme 9-keto-reductase (Weber et al. 1977). However, others observing a much larger change in this ratio have been unable to find a simultaneous change in 9-keto-reductase activity (Walker and Frölich 1981). Furthermore, the activity of this enzyme is so low that it is difficult to see how it can accomplish significant conversion of PGE_2 to $PGF_{2\alpha}$ (Stone and Hart 1975). Recently, Quereshi and Cagen (1982) utilized a double isotope method to determine whether $PGF_{2\alpha}$ produced by slices of rabbit renal medulla or papilla arises from reduction of PGE_2. Comparison of the isotope ratios of the radioactive products with the isotope ratio of the added arachidonic acid indicated that $PGF_{2\alpha}$ was formed by reduction of PGH_2 and not by reduction of PGE_2. Thus, the in vivo function of renal 9-keto-reductase is currently uncertain. There is at present no good evidence for enzymatic formation of $PGF_{2\alpha}$ directly from PGH_2. Because $PGF_{2\alpha}$ shows very low biological activity in most systems tested, it may mainly represent a breakdown product of PGH_2 in aqueous solution which is rapidly eliminated.

Another interconversion which has been reported is the formation of PGE_2 from $PGF_{2\alpha}$ via a 9-hydroxydehydrogenase. Although this enzyme has been found in several tissues, $PGF_{2\alpha}$ itself is a poor substrate in most cases. But Hoult and Moore (1977) reported that in the rabbit kidney 9-hydroxydehydrogenase readily utilizes $PGF_{2\alpha}$. This could be of interest if a quantitatively important amount of $PGF_{2\alpha}$ is converted to the more active PGE_2 in vivo.

1.7 Renal Eicosanoids

In the initial studies on renal eicosanoid synthesis, whole cortex or medulla was studied and the presence of PGE, $PGF_{2\alpha}$, and PGI_2 described (Bergström et al. 1962; Lee et al. 1965, 1969; Larsson and Anggard 1973; Pong and Levine 1976; Whorton et al. 1978a). The earlier studies led to the proposal that PGI_2 is primarily synthesized in renal cortex (in association with vascular elements), while PGE_2 and $PGF_{2\alpha}$ are the major products in the medulla (Frölich et al. 1978; Whorton et al. 1978a; McGiff and Wong 1979). Although this postulate has generally held true, some exceptions have been noted, depending on experimental conditions. For example, 6-keto-$PGF_{1\alpha}$ was found to be a major product in medullary tissues in some studies (Satoh and Satoh 1980; Hassid and Dunn 1982). It is obvious from the complex structure and function of the kidney that such studies can only give a rough estimation of the in vivo situation, as the highly specialized structures of the kidney's vascular and tubular fluid compartment must certainly respond to different synthetic stimuli and with different eicosanoids. In view of the difficulty in interpreting these in vitro results, it is not surprising that often disparate results are reported, depending on differences of cofactors and substrate concentrations, or on whether endogenous or exogenous substrate (arachidonic acid) was utilized. More recent data on renal PG synthesis have been obtained in isolated cells in culture or in microdissected tissue fragments. It has become clear that there is a compartmentalization of PG synthesis, at least with respect to various cell types. Thus, it has been reported that the renal vasculature (McGiff and Wong 1979) and glomeruli (Folkert and Schlondorff 1979; Kreisberg et al. 1982) synthesize predominantly PGI_2, while collecting tubules (Grenier et al. 1981; Schlondorff et al. 1982; Garcia-Perez and Smith 1984) and renomedullary interstitial cells (Zusman and Keiser 1977a; Beck et al. 1980) synthesize predominantly PGE_2 and, to a lesser extent, $PGF_{2\alpha}$. These findings are discussed in more detail in sections which follow.

The renal synthesis of TXA_2 was first discovered in the ureter-obstructed kidney (Morrison et al. 1977) but has also been described in microsomes of normal kidneys (Zenser et al. 1977; Beck and Shaw 1981; Okahara et al. 1983) and in isolated glomeruli (Folkert and Schlondorff 1979; Petrulis et al. 1981). The TX-generating capacity, however, is much lower than for PGs.

Like cyclooxygenase, lipoxygenase can result in the formation of 12- and 15-HETE in glomeruli of rat and man (Jim et al. 1982; Sraer et al. 1983; Winokur and Morrison 1981). Their function is unknown at present. For renal LT synthesis, see Sect. 7.

Furthermore, a P450 mixed-function oxidase can convert arachidonic acid to bis- and trishydroxy fatty acids. The formation of the bishydroxy fatty acids probably involves the formation of epoxides (for example, 11,12 and 14,15 epoxides) via an epoxygenase. The bishydroxy fatty acids are further converted to the trishydroxy fatty acids by w- and w-1 oxidation (Morrison and Pascoe 1981; Mann et al. 1983; Oliw and

Oates 1981). These epoxygenase products may be produced by cells of the thick ascending limb of Henle's loop (Ferreri et al. 1984; Schwartzman et al. 1985; Schlondorff et al. 1982), and the 5,6 epoxide has been shown to inhibit sodium transport (Jacobson et al. 1984). Nothing is as yet known about the in vivo synthesis of these compounds.

In view of the difficulties in interpreting the in vitro data on PG biosynthesis, in vivo measurements have been attempted. Thus, renal arterial and venous PG levels have been measured (Dunn et al. 1978b; Zambraski and Dunn 1979) and a small gradient has been found for PGE_2 and $PGF_{2\alpha}$. However, blood levels of PGs are prone to artifact generation owing to activation of platelets and irritation of the vascular endothelium (Frölich 1979). Also, the technical difficulty and the invasiveness involved in obtaining samples for arteriovenous differences make it impractical or impossible in some situations. Furthermore , it is unclear whether renal venous PGs reflect whole kidney synthesis or only that of some particular region or compartment. We therefore investigated the possibility of a urinary discharge of renally synthesized PGs. In the course of these studies we found that PGE_2, $PGF_{2\alpha}$, and 6-keto-$PGF_{1\alpha}$ are excreted in urine (Frölich et al. 1975c; Rosenkranz et al. 1980; Frölich and Rosenkranz 1980). However, while we found that urinary 6-keto-$PGF_{1\alpha}$ can originate from infusion of it or of PGI_2 (Rosenkranz et al. 1980), PGE_2 and $PGF_{2\alpha}$ do not appear in the urine after intravenous administration (Hamberg and Samuelsson 1971; Granström and Samuelsson 1971). Furthermore, PGE_2 and $PGF_{2\alpha}$ were found to be increased in the ipsilateral urine only when angiotensien II was infused into one renal artery (Frölich et al. 1975e), thus showing that enhanced urinary elimination was due to local renal synthesis. We subseqently determined the site of entry of urinary PGE_2 and $PGF_{2\alpha}$ in the dog and identified it as Henle's loop (Williams et al. 1977). PGE_2 infused into the renal artery is in part secreted by a probenecid-inibitable process (Rosenblatt et al. 1978). Interestingly, however, probenecid has no effect on basal excretion of endogenous PGE_2 (Rosenblatt et al. 1978). This suggests that endogenously synthesized PGE_2 enters the tubuls at a site distal to the organic anion secretory pathways. Excretion of 6-keto-$PGF_{1\alpha}$ infused into the renal artery is not affected by probenecid, thus indicating that small differences in molecular configuration can make significant differences in renal handling of prostanoids (Rosenkranz et al. 1981a). We found that 6-keto-$PGF_{1\alpha}$ is excreted after intravenous infusion (Rosenkranz et al. 1980, 1981a) and thus does not serve as a specific marker of renal synthesis of this prostanoid. Interestingly, Jackson et al. (1984) reported renal arteriovenous differences for 6-keto-$PGF_{1\alpha}$ in dogs with renal artery constriction, indicating that renal PGI_2 biosynthesis may be reflected in the venous efflux of metabolites.

Measurement of urinary PG levels has been found useful for investigations into the regulation of renal PG synthesis. However, we have pointed out some of the limitations of this approach. Seminal fluid has very high concentrations of PGE_2 (50 $\mu g/ml$) and can result in high urinary concentrations of PGE_2 in the male (Reimann et al. 1983; Frölich et al. 1979; Patrono et al. 1979).

Furthermore, PGs are weak acids and consequently their excretion rate can depend on urine flow rate. While there is some evidence that PGE_2 can be excreted flow-dependently (Kaye et al. 1980; Kirschenbaum et al. 1981), there is also clear evidence for the opposite; administration of dDAVP to the Brattleboro rat causes an eightfold reduction in urine flow and a simultaneous sixfold increase in urinary PG excretion (Walker and Frölich 1981). Further studies reveal that the rat, in contrast to man and dog, does not exhibit flow-dependent elimination of PGE_2 (Leyssac and Christensen 1981; Fejes-Toth et al. 1983c). Recent studies have examined the pH dependence of urinary PG excretion. Haylor et al. (1984) reported that PGE_2 excretion is markedly increased in rats by alkalinization of the urine with acute sodium bicarbonate administration and is decreased by an acid load. They suggested that the enhanced ionization of PGE_2 at higher urinary pH results in less reabsorption in the distal nephron. This is consistent with the results of Peterson et al. (1984), who found that bicarbonate loading enhanced urinary recovery of labeled PGE_2 and $PGF_{2\alpha}$ following microinjections into late distal tubules. They also found, however, that even in acidloated rats, reabsorption of the PGs was very low (4%). It is therefore uncertain whether such a mechanism would accound for the larger changes in PGE_2 excretion observed by Haylor et al. (1984). This question clearly deserves further examination.

It is clear from these considerations that we are far from being able, to analyze renal PG synthesis in vivo in ways that permit identification of the site, cell type, and prostanoid involved in a given renal response. Also, virtually nothing is known about markers for estimation of renal synthesis of lipoxygenase products in vivo.

2 Role of Eicosanoids in Regulation of Renal Hemodynamics

2.1 Infusion Experiments Utilizing Eicosanoids or Arachidonic Acid

It has been known for some time that PGE or arachidonic acid, when infused into the renal artery of dogs, causes increased renal blood flow (RBF) (Johnston et al. 1967; Tannenbaum et al. 1975). Other PGs, including PGD_2 and PGI_2, have similar effects (Bolger et al. 1977, 1978; Gerber et al. 1978b, c). Such studies have given rise to the concept that cyclooxygenase products act as vasodilatory agents in vivo. However, discovery of the TXs has complicated this interpretation. TXA_2 is a powerful vasoconstrictor in many vascular beds, and even though it has not been possible to test directly its effect on renal vascular resistance owing to its short half-life, indirect evidence is accumulating which indicates a vascoconstrictor effect in the kidney (see below). LTs have only recently become available for investigation, and their effects on renal hemodynamics have not been widely studied. However, LTC_4 has been shown to increase the glomerular filtration rate (GFR) on systemic infusion in the conscious, unrestrained rat (Filep et al. 1983), and intrarenal infusion of LTC_4 increased RBF in the anesthetized dog (Feigen 1983). Since the vascular elements of the kidney, including glomeruli, have the capacity to form PGI_2, PGE_2, TXA_2, and perhaps LTs (see below), the potential for regulation of RBF or GFR by eicosanoids clearly exists.

One source of controversy with regard to PGs and RBF has been the question of whether PGE_2 shows vasodilatory activity in the rat kidney as in other species. Malik and McGiff (1975) first reported that in the isolated rat kindey PGE_2 causes vasoconstriction rather than vasodilation. In vivo studies were also supportive of this concept (Baer and McGiff 1979; Gerber and Nies 1979). More recently, however, other studies have demonstrated a renal vasodilatory action of PGE_2 (Haylor and Towers 1982; Jackson et al. 1982; Sakr and Dunham 1982; Inokuchi and Malik 1984). All these studies were carried out in vivo and would seem clearly to demonstrate that low doses of PGE_2 have a relaxant effect on the renal vasculature. The reason for the disparate findings is unclear. Some investigators have observed biphasic responses, with increased renal vascular resistance at high doses (Haylor and Towers 1982; Sakr and Durham 1982), but this does not account for all of the discrepancy. Schor and Brenner (1981) observed that both PGE_2 and PGI_2 increased renal vascular resistance in their preparation but that addition of an angiotensin II antagonist converted this in to a dilatory response. Such observations suggest that stimulation of renin release (or intrarenal angiotensin II formation) may explain the vasoconstriction observed in some studies.

Infusion of arachidonic acid has also been shown under some conditions to increase renal vascular resistance in the rat (Gerber and Nies 1979). It is possible that such a response to arachidonate can be explained on the basis of intrarenal TXA_2 formation (Sakr and Durham 1982).

2.2 Inhibition of Cyclooxygenase

A useful approach to the assessment of the role of endogenous PGs in renal hemodynamics is the administration of PG synthetase inhibitors to suppress endogenous synthesis, together with observation of the effects on RBF or GFR. Initial studies, carried out in anesthetized, laparotomized dogs (Lonigro et al. 1973; Aiken and Vane 1973) or perfused kidneys (Herbaczynka-Cedro and Vane 1973; Itskovitz et al. 1973, 1974), demonstrated the profound effects of indomethacin on "basal" RBF and led to proposals that PGs are important regulators of renal vascular resistance under normal conditions. However, in later studies, indomethacin had no effect on RBF in conscious dogs (Swain et al. 1975; Zins 1975; Kirschenbaum and Stein 1976; Fejes-Toth et al. 1978). It seems likely that earlier positive results with indomethacin can be explained by the presence of excessive vasoconstrictor influences arising from the experimental maneuvers (e.g., anesthesia and surgery), which led to RBF dependence on exaggerated PG synthesis. Studies in normal man also indicate no effect of PG synthesis inhibitors on RBF (Donker et al. 1976). A possible exception is the rabbit, since Beilin and Bhattacharya (1977a) and Banks et al. (1980) reported that indomethacin or meclofenamate decreased RBF in conscious rabbits.

2.3 Intrarenal Distribution of Blood Flow

Many studies have also addressed the question of possible effect of PGs on intrarenal blood flow distribution. Intrarenal infusions of PGE_1 or arachidonic acid have been shown to have a more pronounced effect on inner cortical blood flow than on superficial cortical blood flow (Itskovitz et al. 1974; Larsson and Anggard 1974; Chang et al. 1975), and administration of PG synthetase inhibitors preferentially decrease inner cortical flow in the perfused dog kidney (Itskovitz et al. 1973) and in the anesthetized dog (Kirschenbaum et al. 1974). In addition, there is evidence that indomethacin and meclofenamate decrease inner medullary blood flow in the anesthetized rat under some conditions (Solez et al. 1974; Ganguli et al. 1977). Papillary plasma flow was shown to increase after exposure of the renal papilla, while systemic administration of indomethacin or meclofenamate returned the papillary flow to control levels (Chuang et al. 1978). Recently, Lemley and coworkers (1984) observed that erythrocyte velocity in the vasa recta of the rat papilla was decreased after indomethacin pretreatment. Although these findings seem to support the notion of a stimulatory effect of endogenous medullary PGs on papillary blood flow, it is possible that the changes in medullary flow reflect altered hemodynamics in juxtamedullary glomeruli. Stokes (1981) suggested that these glomeruli may have a greater capacity for PG synthesis.

2.4 Importance of Renal Eicosanoids for Regulation of RBF During Stress, Ureteral Obstruction, and Sodium Restriction

The role of PG in the control of intrarenal distribution of blood flow, like, the situation regarding PG effects on total RBF is a matter of study. Thus, it was reported that indomethacin has no effect on intracortical blood flow distribution in conscious dogs (Zins 1975; Fejes-Toth et al. 1978), and it was argued that only with the stress induced by anesthesia and surgery are vasoconstrictor mechanisms activated which create a reliance of juxtamedullary flow on PG synthesis (Zins 1975). The rabbit appears to respond to indomethacin with a redistribution of blood flow, even in the conscious state (Beilin and Bhattacharya 1977a).

Terragno et al. (1977) carried out experiments designed to make direct comparisons of the effects of indomethacin on RBF in groups of conscious, anesthetized, and anesthetized/laparotomized dogs. They also measured renal venous PG efflux and plasma renin activity (PRA) under these conditions. Their results showed that indomethacin did not affect RBF in conscious dogs, even at doses as high as 10 mg/kg intravenously. Furthermore, in anesthetized dogs without surgery, there was little change in PG efflux from the kidney, and indomethacin had little or no effect on RBF. However, in dogs which were anesthetized and actuely operated on, renal PGE efflux was increased eight fold, and indomethacin, in a dose of only 2 mg/kg, decreased RBF by more than 40%. In the three groups there was a highly significant correlation between PGE efflux and PRA, suggesting that PG synthesis is stimulated by the elevated angiotensin levels. The early studies of McGiff and co-workers (1970) support this interpretation, as they showed an increased PG release into renal venous blood during angiotensin II infusion. Studies in the rat (Scherer et al. 1978) also support this hypothesis, since anesthesia and laparotomy induced an eight fold rise in PGE_2 and $PGF_{2\alpha}$ excretion. During anesthesia and surgery, therefore, the elevated angiotensin II levels together with other possible vasoconstrictors probably act to decrease RBF, and the PGs oppose this action. Thus, the reduction of RBF by intrarenal angiotensin II in the conscious dog was significantly enhanced by inhibition of PG biosynthesis (Rosenkranz et al. 1981c). Baylis and Brenner (1978) observed a similar phenomenon in the rat infused with angiotensin II. Satoh and Zimmermann (1975) showed that during renal ischemia (induced by a renal artery clamp) indomethacin markedly decreased RBF. However, infusion of an angiotensin II antagonist abolished the response to indomethacin, thus suggesting that the effect of PGs on RBF was only observable during angiotension-induced vasoconstriction.

It is becoming increasingly clear that under certain other experimental or pathophysiologic states, renal hemodynamics may depend critically on the PG system. It has been demonstrated, for example, that in hemorrhagic hypotension vasoconstrictor influences may reduce renal perfusion, especially in outer cortical regions (Carrière et al. 1966; Logan et al. 1971; Rector et al. 1972), thus causing a redistribution of RBF toward the inner cortex. Data et al. (1976) showed that blockade of PG synthesis prior to hemorrhage inhibits or reverses the redistricution and increases total renal resistance during hypotension. Other investigators have shown similar effects of PG synthesis inhibitors on the whole kidney (Vatner 1974; Tyssebotn and Kirkebo 1977) or on intrarenal distribution responses to hemorrhage (Montgomery et al. 1980). Studies by Henrich et al. (1978a, b) indicate that the renal nerves and the renin-angiotensin system are important mediators of renal vasoconstriction during hemorrhage, and that the PGs oppose their action and in this way maintain renal circulation.

PGs may play a protective role, at least under some conditions, such as ureteral obstruction. Results from various laboratories are difficult to compare because of differences in the degree and duration of ureteral obstruction and species differences. However, some conclusions can already be drawn. The renal hemodynamic response to ureteral obstruction is biphasic, an initial vasodilatation lasting from a few minutes to several hours, followed by a progressive vasoconstriction which results in a marked reduction in RBF (Vaughan et al. 1971; Moody et al. 1975). This increased resistance is maintained for 24 h or longer and persists even after release of the obstruction (Harris and Yarger 1974). The initial phase (vasodilation) of this response can be blocked by PG synthesis inhibitors (Olsen et al. 1976; Allen et al. 1978;

Cadnapaphornchai et al. 1978; Edwards and Suki 1978; Gaudio et al. 1980). These results have been interpreted to suggest that the kidney releases PG in an attempt to increase RBF so to as maintain GFR (Olsen et al. 1976; Gaudio et al. 1980). The role that PGs may play during the second phase (vasoconstriction) has not been so widely studied, but it seems likely that vasodilator PGs may help to oppose the increased resistance. Ichikawa and Brenner (1979) reported that in rats with chronic partial ureteral obstruction, although glomerular plasma flow and filtration rate had recovered to normal levels after 4 weeks, administration of indomethacin or meclofenamate markedly reduced single-nephron GFR and glomerular plasma flow. These changes were confined to the obstructed kidneys. The authors suggested that with partial ureteral obstruction of this duration, PG vasodilation opposes the effects of a vasoconstrictor (probably angiotensin) to maintain glomerular function. An opposing view is that cyclooxygenase products play a causative role in the vasoconstrictor phase. This evidence is discussed further below.

Good evidence is now accumulating that the PGs may help to preserve renal perfusion during sodium restriction. Circulating angiotensin II, a potent renal vasoconstrictor, is elevated during sodium depletion, and earlier studies have shown that administration of angiotensin antagonists under these conditions markedly elevates RBF (Mimran et al. 1975; Lohmeier et al. 1977). However, RBF is generally not reduced during sodium depletion in intact animals, thus suggesting that one or more factors oppose the effect of angiotensin. Several recent studies demonstrate that in the anesthetized, sodium-depleted dog (Blasingham and Nasjletti 1980; Oliver et al. 1980) and rat (Schor et al. 1980) administration of cyclooxygenase inhibitors markedly increases renal vascular resistance and decreases RBF and/or GFR. Similar changes were not observed in sodium-replete animals. These findings were also reported for conscious dogs (Blasingham et al. 1980; DeForrest et al. 1980) and in human volunteers (Rosenkranz et al. 1981c; Muther et al. 1981). The studies carried out in the rat are of interest because the problem was examined at the glomerular level (Schor et al. 1980). The latter authors conclude that the changes in single-nephron GFR and plasma flow caused by indomethacin in low-salt animals were due to increases in both efferent and afferent arteriolar resistances. The PGs therefore, seem to be important in the maintenance of RBF and GFR during sodium depletion, when angiotensin or other vasoconstrictors (e.g., norepinephrine; (Oliver et al. 1980) threaten renal perfusion. In this regard, it should be mentioned that the effect of sodium intake on renal PG synthesis has been an area of considerable controversy (see Sect. 3 below). However, there have been reports of increased PGE_2

excretion in the rabbit during low salt intake (Weber et al. 1979). Furthermore, a recent study (Stahl and Attallah 1980) indicates that treatment of sodium-depleted rabbits with saralasin, a competitive antagonist of angiotensin II, decreases PG excretion, a finding consistent with the suggestion that angiotensin acts to increase renal PG synthesis. In fact, studies of the effect of angiotensin II on urinary PGE_2 excretion in man have shown a dose-dependent increase (Frölich et al. 1975c). Oliver et al. (1980) also report that PGE_2 secretion by the kidney of the sodium-depleted dog is enhanced, as is the secretion of renin and norepinephrine.

2.5   Renal Eicosanoids in Regulation of RBF in Human Disease States

As might be expected from these studies in various experimental settings, a role for PGs in the maintenance of RBF and GFR in disease states in man has been considered. As pointed out by Epstein and Lifschitz (1980), this is well documented in those disorders classically categorized as states of low effective plasma volume; cirrhosis with ascites, congestive heart failure, and nephrotic syndrome. Arisz et al. (1976) reported that in patients with the nephrotic syndrome, indomethacin treatment caused a striking decrease in GFR, and this decrease was more prominent during sodium restriction. In patients with congestive heart failure (Walsche and Venuto 1979), urinary PGE excretion was reported to be markedly elevated, and in one patient given indomethacin, GFR was reduced and serum creatinine increased, thus suggesting that the elevated renal PG synthesis helps to compensate for renal vasoconstrictor stimuli in heart failure.

In support of this hypothesis, Oliver et al. (1981) recently reported studies in dogs in which a reduction in cardiac output was accomplished by inflation of a balloon in the vena cava. In this model of "heart failure", the secretion of renin, norepinephrine, and PGE by the kidney was enhanced, but RBF was well maintained. With the addition of indomethacin, renal vascular resistance increased markedly and RBF was reduced. Also of interest with regard to congestive heart failure and renal PGs are studies in infants with a patent ductus arteriosus. For several years indomethacin has been used to accomplish closure of the ductus, presumably by inhibition of vascular PG synthesis. Many reports of elevations of blood urea nitrogen and serum creatinine during indomethacin administration have appeared, and recent studies (Cifuentes et al. 1979; Catterton et al. 1980; Halliday et al. 1979) document the decrease in GFR with indomethacin. It is, however, uncertain whether this decrease in GFR relates to the decompensated cardiac status or to perhaps in complete nephrogenesis in these premature infants (Cifuentes et al. 1979).

With regard to liver disease, three recent studies (Boyer et al. 1979; Zipser et al. 1979; Arroyo et al. 1983) indicate that PG synthesis in-inhibitors can induce marked deterioration of renal function (GFR) in cirrhotic patients with ascites, especially in those with avid sodium retention. Zipser et al. (1979) and Arroyo et al. (1983) also report elevated PGE excretion in these patients. Epstein et al. (1982) did not observe higher urinary PGE levels in decompensated cirrhotics, but recent studies in the dog are in agreement with the former reports. In fact, in two reports of experimentally induced hepatic cirrhosis in dogs, indomethacin markedly reduced RBF and GFR (Levy et al. 1983; Zambraski and Dunn 1984). Taken together, all these results in diseases of depleted effective blood volume suggest that PGs play an important role in modulation of renal resistance in the face of elevated circulating vasoconstrictors.

Other studies show that renal PGs may be critical in other pathologic states where renal function is compromised or threatened. Thus, in Bartter's syndrome (Bowden et al. 1978; Gill et al. 1976), chronic renal insufficiency (Berg 1977), systemic lupus erythematosus with renal involvement (Kimberley and Plotz 1977; Kimberley et al. 1978), and various other renal abnormalities (Donker et al. 1976; Ciabattoni et al. 1984), administration of PG synthesis inhibitors resulted in acute de-creases in GFR. This acute renal failure may also occur in patients with apparently normal renal function, but its incidence appears to be much lower. As physicians become aware of this problem, it seems likely that more reports documenting this potential danger of non steroidal anti-inflammatory drugs will appear (for review see Clive and Stoff 1984).

2.6 Renal Autoregulation

The marked dependence of renal perfusion on continued PG synthesis under many experimental conditions led to proposals that PGs may be responsible for adjustments in renal resistance to maintain RBF and/or GFR during changes in perfusion pressure over the so-called autoregulatory range (approximately 80–180 mmHg in most species). This phenomenon was described many years ago, and there is still controversy as to the mechanisms of the adjustments. Early reports that PG synthesis in-hibitors impair autoregulation of RBF (Herbaczynka-Cedro and Vane 1973) have not generally been supported by more recent data (Anderson et al. 1975; Owen et al. 1975; Venuto et al. 1975; Finn and Arendshorst 1976; Kaloyanides et al. 1976; Beilin and Bhattacharya 1977b).

However, Schnermann et al. (1981, 1984) examined this question in detail in the rat, measuring single-nephron GFR. They presented evidence that PGs are involved in the adjustments in GFR to lowered renal perfusion pressure, especially in the lower autoregulatory range of arterial blood pressure (75 to 95 mmHg). In this pressure range, indomethacin appears to reduce autoregulatory efficiency, partly by interference with the tubuloglomerular feedback response (Schnerman et al. 1979). It is uncertain how these results may be reconciled with studies finding no effect of indomethacin on autoregulation. One possible explanation is that the PGs may be more important for GFR autoregulation than for RBF, since previous studies in the rat (Finn and Arendshorst 1976) did not examine GFR. There is no direct evidence for this, however, and studies in the dog (Anderson et al. 1975) showed no effect of indomethacin or meclofenamate on autoregulation of GFR.

2.7 Eicosanoids as Renal Vasoconstrictors

In addition to the evidence which implicates cyclooxygenase products in a vasodilator role in the kidney, some studies have been suggestive of a constrictore effect under certain conditions, presumably mediated by TXA_2. Morrison et al. (1977, 1978) demonstrated that the ureter-obstructed rabbit kidney, in contrast to the contralateral or normal kidney, has the capacity to synthesize appreciable amounts of TXA_2, a powerful vasoconstrictor. It has been suggested that this substance may mediate the increased resistance during prolonged ureteral obstruction. Recent evidence is consistent with this suggestion. Yarger et al. (1980) examined renal function in the rat immediately after release of 24 h unilateral ureteral obstruction. A severe vasoconstriction was observed in the postobstructive kidney. Indomethacin, which would inhibit both PG and TX biosynthesis, did not improve renal hemodynamics. However, large doses of imidazole, which inhibits TX synthesis fairly selectively (Needleman et al. 1977), effected a marked improvement in filtration rate and RBF. Whinnery and coworkers (1982) showed that in renal slices from obstructed rat kidneys, TXB_2 production is enhanced, even in the absence of hormonal stimulation by bradykinin or angiotensin.

Findings similar to these in the ureter-obstructed kidney have been obtained in other pathologic conditions. Zipser et al. (1980) recently reported that in the rabbit with chronic renal venous constriction exag-

gerated PG biosynthesis occurs in response to bradykinin and angiotensin II, and TX biosynthesis is unmasked. Other investigators have
evaluated the PG system in glycerol-induced acute renal failure. Benabe
et al. (1980) provided evidence that the kidneys of glycerol-treated
rabbits are also able to synthesize TX when perfused ex vivo, and a good
correlation was obtained between the elevated serum creatinine and the
percent conversion of arachidonic acid to TXB_2. Recently published
studies (Okegawa et al. 1983; Schwartz et al. 1984). indicate that in the
ureter-obstructed or renal vein-constricted rabbit kidney, infiltration
of damaged tissue with mononuclear cells may be responsible for, or at
least contribute to, the enhanced TXB_2 production. It is clear that
this question of interaction of various cell types in PG and TX production in inflamed organs deserves further study.

Other studies have implicated TXA_2 in a renal vasoconstrictore role.
Beck and Shaw (1981) demonstrated that the kidneys of potassium-
depleted rats synthesize increased amounts of TXB_2 in vitro, as compared
to controls. Linas and Dickmann (1982) observed a reduced RBF in
potassium-depleted conscious rats and found that administration of
either meclofenamate or imidazole could partially restore perfusion to
normal. Angiotensin II also seemed to contribute to the vasoconstriction.

In view of the complex interactions of PGs with other vasoactive
factors (e.g., catecholamines, kallikrein-kinin system), it seems likely
that much of the contradiction will be eliminated by future studies in
which these other factors are controlled or accounted for.

3 Eicosanoids and Renal Electrolyte Excretion

3.1 Sodium

No area of renal PG research is more controversial than that of the
effects of PGs on sodium excretion (Kirschenbaum and Serros 1980).
Although a natriuretic effect is readily demonstrable upon intrarenal
infusion of PGE or arachidonic acid (Johnston et al. 1967; Strandhoy
et al. 1974; Tannenbaum et al. 1975), this approach seems unlikely
to provide firm evidence as to a physiologic role for PGs in sodium
handling. The application of PGs (or arachidonic acid) by this route is
different from localized synthesis of PGs intrarenally. In addition, the
renal vasodilation observed during such infusions may increase sodium
excretion in a non specific manner (Martinez-Maldonado et al. 1972),
since other vasodilators can also increase sodium excretion, presumably

via effects on physical factors influencing proximal salt and water reabsorption.

In attempts to identify a possible relationship between endogenous renal PGs and sodium balance, many investigators have evaluated the effects of altering sodium intake on renal PG synthesis. These studies have not resulted in any clarification of the tissue. Several studies have shown reciprocal changes in renal excretion or secretion of PGE_2 and dietary sodium, thus suggesting an antinatriuretic rather than natriuretic effect of PGs in rabbits (Scherer et al. 1977; Davila et al. 1978) and dogs (Oliver et al. 1980). However, it appears that these changes in PG synthesis are of little consequence of sodium handling, since administration of cyclooxygenase inhibitors in the rabbit apparently does not alter sodium excretion (Davila et al. 1980; Lifschitz et al. 1978). In addition, other investigators have observed increases in PG excretion with high salt intake in man (Kaye et al. 1979) and the rat (Tan et al. 1980), which suggest a natriuretic role, although in the latter study the changes were transient. However, cyclooxygenase inhibitors were also without effect on sodium excretion under these conditions. Thus, there is no convincing evidence from these types of studies that PGs play a role in sodium handling. It seem probable that the factors which influence PG excretion are complex and under most conditions cannot be related directly to a single variable such as sodium intake. In addition, methological problems in the assay of PGs may be an important factor in the confusion.

The approach most often used in the assessment of naturetic or antinatriuretic roles for PGs has been the administration of inhibitors of PG synthetase. The results obtained in these studies are somewhat more consistent, although not without conflict. Various investigators have found acute reductions in sodium excretion after administration of indomethacin or other nonsteroidal anti-inflammatory agents in the normal rat (Feldman et al. 1978; Leyssac et al. 1975; Higashihara et al. 1979; Kadokawa et al. 1979; Haylor and Lote 1980), dog (Fejes-Toh et al. 1977; Blashingham and Nasjletti 1980; Olsen et al. 1976; Feigen et al. 1976; Williamson et al. 1978), and man (Frölich et al. 1976c, 1978; Brater 1979; Donker et al. 1976; Berg et al. 1977; Haylor 1980). Other studies have demonstrated inhibition of the natriuretic responses to extracellular fluid volume expansion in the rat (Düsing et al. 1977; Higashira et al. 1979; Susic and Sparks 1975) and to central volume expansion in sodium-depleted man (Epstein et al. 1979).

Some support is also provided for a natriuretic role for PGs in micropuncture studies in the rat. Leyssac et al. (1975) and Roman and Kauker (1978) demonstrated enhanced sodium reabsorption at a point beyond the late proximal tubule after indomethacin or meclofenamate administration in rats. Higashihara et al. (1979) obtained similar results for chloride, showing that in hydropenic rats, meclofenamate increased chloride reabsorption beyond the late distal tubule and during volume expansion in the thick ascending limb as well. In a related study, Kauker (1977) demonstrated that the urinary recovery of ^{22}Na, microinfused into late proximal tubular puncture sites, was slighthly increased by addition of PGE_2 to the perfusate. Although very high concentrations of PGE_2 were utilized in the latter study, the consistent conclusion from these micropuncture investigations is that sodium reabsorption is inhibited by PGs in some segment beyond the proximal tubule, most likely in the collecting system or in the ascending limbs of Henle's loop. It is to be remembered that in conventional micropuncture studies, the investigator samples only the surface nephron population, and extrapolation to deeper nephrons may not be valid. For example, an increase in the fraction of filtered sodium remaining between late distal puncture sites and urine may reflect collecting duct secretion or may indicate that deeper (juxtamedullary) nephrons deliver a much larger fraction of filtered sodium to urine than do superficial ones. In fact, several recent studies have produced evidence that this may be the case in volume expansion, i.e., in the saliuresis of volume expansion, juxtamedullary nephrons reabsorb proportionally less sodium chloride than do superficial ones (Osgood et al. 1978; Higashihara et al. 1979; for review see Walker and Valtin 1982) and, therefore, account for an increase in the fraction remaining between late distal sites and urine. These considerations may be of special importance to the renal PG system, since Higashihara and coworkers (1979) were able to show by cortical and papillary micropuncture techniques that indomethacin and meclofenamate inhibit this contribution of juxtamedullary nephrons, i.e., chloride reabsorption in this population is preferentially enhanced by PG synthesis inhibition.

   This latter report also confirmed a previous interesting finding that papillary chloride concentrations are increased by cyclooxygenase inhibitors. It had been reported by Bartelheimer and Senft (1968) that oxyphenbutazone increased the corticomedullary sodium concentration gradient in rats, and later Ganguli et al. (1977) and Haylor and Lote (1983) demonstrated an increase in papillary sodium and chloride concentration in rats after administration of indomethacin or meclofenamate. The available evidence suggests that these changes are not dependent on changes in medullary blood flow (Ganguli et al. 1977), although the methods for measuring blood flow in this portion of the kidney are probably inadequate to assess subtle changes. At any rate, the marked accumulation of salt in the papillary interstitium is certainly further strong evidence that PGs are in some way involved in electrolyte handling.

Studies in isolated renal tubules have provided some support for PGs as modulators of sodium or chloride transport. Stokes and Kokko (1977) published evidence which demonstrated that PGE_2 inhibits sodium efflux from isolated, perfused rabbit cortical collecting tubules without affecting backflux, thus suggesting direct or indirect inhibition of active sodium transport. These results were confirmed in another laboratory (Iino and Imai 1978). Effects were seen at concentrations of PGE_2 as low as 10^{-7} M and were reversible. Stokes (1979) carried out similar investigations in thick ascending limbs, noting inhibitory effects of PGE_2 on chloride transport in medullary but not cortical segments.

Of interest with regard to the studies of thick ascending limbs is the recent surge of interest in the effects of antidiuretic hormone (ADH) in this portion of the nephron of small rodents (Walker and Valtin 1982; Hebert and Andreoli 1984). Several reports demonstrate a stimulatory effect of this peptide on sodium chloride transport in thick ascending limbs (Hall and Varney 1980; Sasaki and Imai 1980; Culpepper and Andreoli 1983). This action of ADH increases the addition of solute to the medullary interstitium and thereby enhances the buildup of the corticomedullary osmotic gradient. Culpepper and Andreoli (1983) examined the interaction of ADH and PGE_2 in isolated medullary thick ascending limbs of mice. They concluded that ADH stimulated sodium and chloride transport, and that PGE_2 inhibited this increase in transport without altering the ADH-independent component. These studies are of interest because the effects of ADH and PGE_2 were observable at relatively low concentrations of both (apparent Ka's of of 3 pM and 100 pM respectively). Endogenous PGs may therefore inhibit sodium transport in the medullary thick ascending limb. In vivo studies are consistent with this proposal in that nonsteroidal anti-inflammatory drugs appear to enhance electrolyte reabsorption in the thick ascending limb (Kaojarern et al. 1983).

ADH and PGE_2 interactions on sodium transport have also been studied in rabbit cortical collecting tubules perfused in vitro (Holt and Lechene 1981a). In this segment ADH caused a sharp increase in sodium transport, followed by a sustained inhibition. This inhibition was prevented by pretreatment of the tubules with meclofenamate. Exogenous PGE_2 mimicked the ADH-induced inhibition of sodium transport. These results suggest that at least at fairly high concentrations, ADH increases PG synthesis in the cortical collecting tubule and that the PGs inhibit sodium transport. In keeping with this conclusion, Fejes-

Toth et al. (1977) observed that PG synthesis inhibitors block the natriuretic response to ADH in hydropenic dogs. This nephron segment, then, is another locus of potential interaction between ADH and PGs (see Sect. 5 below) and one that may have relevance to salt excretion.

The previously cited experiments provide an impressive array of evidence that renal PGs may modulate sodium and chloride reabsorption. However, the sodium-retaining effects of PG synthetase inhibitors have not been demonstrable under many conditions. For example, although there are exceptions (Fejes-Toth et al. 1977; Haylor 1980), an antinatriuretic response has not generally been observed in animals undergoing a water diuresis (Kirschenbaum and Stein 1976; Berl et al. 1977; Altsheler et al. 1978; Gagnon and Felipe 1979; Work et al. 1980; Lameire et al. 1980). But others have also reported no effect during hydropenia (Düsing et al. 1977; Mountokalakis et al. 1978; Patak et al. 1975). Although many investigators have found that PG synthesis inhibitors blunt the natriuretic responses to saline loading (see above), this is not a universal result (Lameire et al. 1980; Kaye et al. 1979; Tan et al. 1979). In fact, there are at least two published studies in which indomethacin enhanced the natriuretic response to volume expansion in the rabbit (Oliw et al. 1978) and in man (Mountokalakis et al. 1978). The above cited study of Kirschenbaum and Stein (1976) also demonstrated a natriuresis due to cyclooxygenase inhibitors during water loading. Thus, with regard to the effects of nonsteroidal anti-inflammatory drugs on solium excretion, there is no consensus under any experimental condition tested.

The effects of PGs on electrolyte transport in isolated tubules (Stokes and Kokko 1977; Iino and Imai 1978; Stokes 1979) have also been questioned by other workers. Fine and Trizna (1977) did not observe any effect of PGE_2, even at very high doses, on sodium transport in either collecting tubules or thick ascending limbs perfused in vitro. The issue has not yet been resolved, and the reader is referred to recent reviews (Kokko 1981; Fine and Kirschenbaum 1981; Stokes 1981) fore more detailed discussions of these discrepancies.

The bulk of the evidence stems from studies utilizing PG synthesis inhibitors. It seems impossible to reconcile the disparate results. Most obvious explanations, such as species differences or experimental design, do not seem to account for the conflict. However, it appears that several points warrant consideration for design of future experiments and for interpretation of available data. First, as mentioned in Sect. 2, under conditions where vasoconstrictor systems are activated, indomethacin and other inhibitors can have marked effects on renal hemodynamics. Therefore, studies in anesthetized/laparotomized animals or in other experimental situations where maintenance of renal hemodynamics

is likely dependent on PG synthesis stress can significantly affect the out-
come of studies of PG synthesis inhibitors on sodium excretion. Second,
careful attention to the previous volume status of the experimental
subjects is also imperative (Epstein and Lifschitz 1980). This is true
because sodium depletion, with attendant activation of the renin-
angiotensin system, is another condition which may create a reliance of
renal perfusion on PGs (Blasingham and Nasjletti 1980). In addition,
although sodium depletion may make the antinatriuretic effects of
indomethacin more difficult to demonstrate (Frölich et al. 1978; Blasingham and Nasjletti 1980), the blunting by indomethacin of the natriuretic response to acute volume expansion is enhanced in sodium-depleted
subjects (Epstein et al. 1979). The prior state of sodium balance may
therefore be critical to the evaluation of roles for endogenous PGs.
Third, the specificity of the drugs sould be kept in mind. The effects
of cyclooxygenase inhibitors in most experiments are not confined
to the kidney, i.e., systemic or central nervous system effects or neural
or humoral influences on renal function could be involved. Also, effects
of cyclooxygenate inhibitors on other enzyme systems have been
demonstrated for many of these drugs (Flower 1974). Fourth, the
duration of treatment with cyclooxygenase inhibitors may be an important consideration. It seems certain that any acute effects of those
drugs which alter sodium excretion will initiate compensatory adjustments
to maintain homeostasis. This is a possible explanation for the apparent
absence of effects of indomethacin on sodium excretion in some chronic
studies (Davila et al. 1980; Lifschitz et al. 1978; Mountokalakis et al.
1978). Kadokawa et al. (1979) observed in rats diminished sodium-
retaining effects of anti-inflammatory drugs with repeated administration.
Fifth, the effectiveness of the inhibitors in acting on PG synthesis
should be established. It is unsafe to assume that because a certain dose
gives 50% suppression of PG synthesis in one experimental situation,
it will be just as effective in another. Several recent reports suggest
that significant stimulation of PG synthesis by various manipulations
can occur even in the presence of doses of indomethacin, which markedly
suppress basal synthesis. Thus, in the study by Epstein et al. (1979),
although indomethacin inhibited urinary PGE excretion by 50% or
more in the control state, a definite increase in PG excretion occurred
during water immersion. Similarly, PG excretion has been shown to
increase with furosemide administration in proportion to the log of
the urinary furosemide concentration (Brater et al. 1980). Although
indomethacin lowered the absolute amount of PGE_2 excreted, the
relative increment induced by furosemide was not affected. In situations
such as these, if not effect of indomethacin on sodium excretion is

observed, one cannot conclude that PGs are not involved; the increment in PG synthesis, though smaller in absolute terms, may be sufficient to play a role even in the presence of indomethacin.

Finally, another consideration may have relevance for the conflicting results obtained with cyclooxygenase inhibitors. There seems to be no good evidence that PGs affect the active transport step in sodium re-absorption. Kokko (1981) proposed, in discussing results of isolated tubule studies, that the inhibitory effects of PGE_2 on sodium transport in the cortical collecting tubule may be due to a decrease in the permeability of the luminal membrane to the passive entry of sodium. If changes in passive permeability characteristics are responsible for the natriuretic effects of PGs, some of the seemingly conflicting data may be explained. Passive movement will of course depend in part on concentration gradients which exist between the tubular lumen and interstitium. Furthermore, these gradients may be altered or reversed, depending on experimental conditions. For example, it is likely that during water diuresis the concentration of sodium in collecting tubule fluid is much lower relative to the interstitium than in hydropenia. This may account for the failure of indomethacin to increase sodium reabsorption in many studies in water-loaded animals. Also, in volume-expanded animals, the normal gradient favoring passive sodium reabsorption between thin ascending limbs and interstitium is reduced (Osgood et al. 1978). If this occurs during chronic expansion, it may explain the reduced effectiveness of indomethacin in enhancing sodium reabsorption during acute volume expansion in salt-replete subjects (Epstein et al. 1979) or with repeated administration of indomethacin (Kadokawa et al. 1978; Mountokalakis et al. 1978).

3.2 Electrolytes Other than Sodium

In addition to sodium, other electrolytes may also be influenced by PG synthesis inhibitors. Lithium plasma levels have been found to increase with the administration of indomethacin, phenylbutazone, ibuprofen, and diclofenac (Imbs et al. 1979; Frölich et al. 1979c; Ragheb et al. 1980; Reimann and Frölich 1981). This effect is due to a reduction in renal lithium clearance. Interestingly, aspirin, in a dose sufficient to cause inhibition of renal PGE_2 synthesis, had no effect on steady state plasma levels or clearance of lithium (Reimann et al. 1983a). These findings indicate that there are important differences among various drugs which inhibit PG synthesis and that results obtained with one drug do not allow extrapolation to the entire "class" of drugs.

Potassium excretion may also be affected by the nonsteroidal anti-inflammatory drugs. This is generally not observed in acute studies, but hyperkalemia has often been reported as a result of prolonged therapy with these agents (for review see Clive and Stoff 1984). The hyperkalemia may be partially attributed to reduced GFR in some patients, but the principal mechanism is thought to be via reduction in PRA and consequent hypoaldosteronism.

4 Role of Eicosanoids in Renal Renin Release

There is a large body of data indicating that PGs stimulate the release of renin. The evidence rests on infusion experiments with PGs or the PG precursor, arachidonic acid, and on administration of PG synthesis inhibitors. Furthermore, in vitro experiments utilizing renal cortical slices, isolated glomeruli, or isolated perfused kidneys support the concept that PGs stimulate renin release.

4.1 Experiments Utilizing Prostanoids or Arachidonic Acid

Studies infusing PGs (PGE_1, PGE_2, PGA_1, PGI_2, 13, 14 dihydro-PGE_2) were equivocal but generally suggestive of a stimulatory effect on renin release as a consequence of vasodilatation (Vander 1968; Werning et al. 1971; Fichman et al. 1972; Krakoff et al. 1973; Gerber et al. 1978b, c, 1979; Patrono et al. 1982), even though there were differences with respect to the magnitude of effects on vasodilation and renin release among these compounds. This approach, however, yields limited information regarding physiologic roles since a nonspecific component, such as vasodilation, may account for the renin release. Furthermore, as mentioned above, the intrarenal infusion of PGs is not an appropriate way to mimic the endogenously synthesizd PGs.

Infusion of arachidonate brings to the enzyme system substrate capable of PG synthesis. In experiments employing rabbits, rats, or dogs, infusion of arachidonate resulted in renin release which could be blocked by indomethacin (Larsson et al. 1974; Weber et al. 1975; Bolger et al. 1976; Gerber et al. 1978a). Furthermore, arachidonate was found to stimulate renin release in renal cortical slices (Weber et al. 1976). These results are more suggestive of possible physiologic roles, as the provision of the precursor allows for synthesis at the location of the cyclooxygenase enzyme, but these studies have also been criticized for the reasons given in Sect. 2.

4.2 Inhibitor of Cyclooxygenase

A more useful approach to the study of PG/renin interactions has been by use of PG synthesis inhibitors. Such studies were based on early reports of a reduction of PRA by indomethacin in rabbits (Larsson et al. 1974; Romero et al. 1976) and in man (Frölich et al. 1976c). The questions that have been addressed in the most recent studies are how PG synthesis inhibitors reduce renin release, the mechanisms of renin release in which PGs may play a role, and the identification of the PGs involved (Gerber et al. 1981a; Henrich 1981; Frölich 1981).

The early observation of a renin-suppressing effect of indomethacin in man immediately highlighted the problems inherent in the use of this drug as a tool in the exploration of a prostanoid mechanism of action: indomethacin not only lowered PRA but also caused sodium retention (Frölich et al. 1976c). The contribution of sodium retention to the reduction in PRA was investigated by placing normal volunteers on a very low sodium diet, followed by administration of indomethacin. Indomethacin lowered PG synthesis and had no effect on sodium balance; however, it failed to reduce PRA (Frölich et al. 1976c). This finding prompted the investigators to consider the possibility that the increased catecholamine levels known to occur during sodium depletion (Gordon et al. 1967) might have caused renin stimulation that was independent of cyclooxygenase. Therefore, the study was repeated after pretreatment with propranolol to exclude the renin stimulation by catecholamines during sodium depletion. Under these circumstances, indomethacin causes a marked and reversible reduction of PRA in the supine and upright postures (Frölich et al. 1979b). This study, therefore, unequivocally showed that in man renin release following cyclooxygenase inhibition is reduced independently of sodium retention. It also suggested that catecholamine-induced renin release is independent of PGs.

This latter possibility was explored by infusing isoproterenol into normal volunteers. The increase in PRA observed was unaffected by indomethacin, thus showing that beta-receptors-mediated renin release is independent of PGs (Frölich et al. 1979b). This finding was questioned on the basis of other studies which showed that PG synthesis inhibition prevented isoproterenol-stimulated renin release (Campbell et al. 1979; Feuerstein and Feuerstein 1980; Suzuki et al. 1981). It seems likely that in at least some of these studies, the failure to control for other mechanisms of renin release (i.e., vascular or macula densa-mediated) may account for the discrepancy. Other studies, some of which provided for elimination of these variables confirmed that beta-receptor-mediated renin release was independent of the PG system (Berl et al. 1979; Seymour et al. 1982; Kopp et al. 1981). Work in vitro systems also shows that the beta-adrenergic response does not depend on PG synthesis. (Beierwaltes et al. 1980; Barchowsky et al. 1984; Linas 1984).

4.3 Alpha- and Beta-Receptor-Mediated Renin Release

Controversy exists with respect to the contribution of alpha-receptors to the release of renin. Phenylephrine can cause enhanced renin release and this increment appears to be PG-dependent (Olson et al. 1981; Vikse et al. 1984). The increase in renin release may be mediated by a mechanism involving sodium transport at the macula densa (Olson et al. 1981) or afferent arteriolar dilation (Vikse et al. 1984). Kopp and coworkers (1981) obtained similar results with renal nerve stimulation in dogs, i.e., PGs were not involved in the beta-receptor-mediated component but played a role in the renin release attributable to alpha-receptor stimulation.

The question of what relative role alpha- and beta-receptor stimulation plays in the release of renin controlled by endogenous catecholamines is at present uncertain. Norepinephrine can stimulate both alpha- and beta-receptors in man and dog. Because the studies on sodium-depleted subjects showed that the beta blocker, propranolol, was very effective in permitting expression of the PG-mediated renin control mechanisms (Frölich et al. 1979), it would appear that at least under conditions of sodium depletion, endogenous catecholamines enhance renin release predominantly by beta-receptor stimulation.

4.4 Relation of Eicosanoid-Mediated Renin Release
to Basic Release Mechanisms

Systematic studies on the mechanisms of renin release in which PGs participate have to consider the relevant current theories (Davis and Freeman 1970). The three basic mechanisms increasing renin release are thought to be: adrenergic stimulation of the renin-releasing cells,; a decrease in sodium delivery at the macula densa; and a decrease in arterial pressure or stretch sensed by a baroreceptor (Davis and Freeman 1970). All stimuli to renin release are thought to be explicable by one of these three basic mechanisms or their combination. The investigative models studied are almost exclusively in the dog. To eliminate the renal baroreceptor, the kidney is maximally vasodilated by papaverine. Any increase in release above that seen with papaverine is then thought to be mediated by one of the other two mechanisms. The macula densa is eliminated by rendering the kidney nonfiltering (Blaine et al. 1970). This is accomplished by clamping the renal artery; opening of the clamp results in deposition of protein precipitates in the tubules and a fall in GFR of over 90%, and a severe reduction in RBF. The catecholamine-induced renin release is blocked by renal denervation, while some investigators also perform adrenalectomy and propranolol infusion in order to prevent the effects of circulating catecholamines (Data et al. 1978).

Studies utilizing these techniques have shown that during propranolol administration in adrenalectomized dogs with a single denervated non-filtering kidney, mechanical reduction of RBF leads to a large increase in renin release. Because the beta- and macula densa mechanisms were eliminated in this model, the response could only be due to activation of the baroreceptor. Administration of indomethacin completely abolished the increase in renin, thus demonstrating that the baroreceptor depends on a cyclooxygenase product for its function (Data et al. 1978). This conclusion was also reached by others using various in vivo dog preparations (Blackshear et al. 1979; Henrich et al. 1979; Berl et al. 1979), although there is some controversy as to whether or not this always holds true for severe reductions in renal perfusion (below the autoregulatory range) (Blackshear et al. 1979; Seymour and Zehr 1979; Freeman et al. 1982). Other workers, using a rat preparation with a single denervated, nonfiltering kidney, did not observe an effect of meclofenamate on the renin release response to either mild or severe reductions in renal perfusion pressure (Villarreal et al. 1984). The explanation of these discrepancies is unclear at present. An elegant study by Linas (1984) clearly showed the PG dependence of the baro-receptor mechanisms in the isolated perfused rat kidney.

The macula densa renin release mechanism has also been investigated with respect to PG involvement. Stimulation of this mechanism by reducing sodium delivery to the macula densa through reduction of renal perfusion pressure resulted in increased renal venous PRA which was blocked by indomethacin or meclofenamate (Gerber et al. 1981b). In these studies, papaverine and propranolol were used to isolate the macula densa mechanism from the baroreceptor and beta-adrenergic influences. Francisco et al. (1982) came to similar conclusions from their in vivo studies, and Linas (1984) demonstrated by three separate approaches the PG dependence of the macula densa renin release mechanism in the isolated perfused kidney. Evaluation of the effects of furosemide on renin release has also lent support to the concept that the macula densa mechanism utilizes a PG pathway. It is well known that furosemide and bumetanide increase RBF and renin release, these increases being accompanied by an increase in renal PG biosynthesis (Ludens et al. 1968; Bailie et al. 1973; Olsen and Ahnfelt-Ronne 1976; Williamson et al. 1975). Several investigators have shown that inhibition of PG biosynthesis blocks the vasodilator response as well as the increase in renin release following administration of furosemide (Bailie et al. 1975; Williamson et al. 1975; Frölich et al. 1976c; Romero et al. 1976; Olsen and Ahnfelt-Ronne 1976; Weber et al. 1977b). One unifying scheme which would account for these observations is based on the knowledge that furosemide decreases ion transport at the macula densa and thereby interrupts the tubuloglomerular feedback mechanism, with consequent afferent arteriolar dilation (Wright and Schnermann 1974; Briggs and Wright 1979). Gerber and Nies (1980) proposed that the renal vascular response to furosemide is the result of a tubular (macula densa) mechanism and is mediated by a PG. In support of this contention, they showed that in the nonfiltering kidney, the renal vascular response to furosemide is absent. Furthermore, desoxycorticosterone acetate treatment, which blunts the tubuloglomerular feedback mechanism, also blunts the renal vascular response to furosemide. These findings suggest that a reduced ion transport at the macula densa results in increased PG biosynthesis, which in turn leads to renin release and increased RBF.

A very carefully conducted study (Seymour and Zehr 1979) blocked all three mechanisms of renin release by utilizing a dog model with a single nonfiltering, denervated kidney under the influence of papaverine and measured renal renin secretion. In this study, arachidonic acid enhanced renin secretion fourfold. However, indomethacin failed to have any effect on this response, in spite of its measured reduction of PG biosynthesis. This finding suggests that arachidonic acid can have an effect on renin release independent of PG synthesis. The conclusion has been drawn that this study shows a direct stimulatory effect of arachidonic acid on the renin-releasing cell, invoking a novel mechanism of renin release which would be additional to the three proposed mechanisms. An effect of arachidonic acid independent of PG synthesis is possible and biochemical pathways exist for the generation of biologically active metabolites, especially via the lipoxygenase pathway (see section on biochemistry of PGs). This possibility has not been explored. However, other investigators utilizing a very similar model, differing only by virtue of a lack of papaverine infusion, have not been able to demonstrate an effect of arachidonic acid on renin release independent of PG synthesis (Weber et al. 1976). Furthermore, indomethacin and other nonsteroidal anti-inflammatory drugs are thought to act by competitive displacement of arachidonic acid from the cyclooxygenase pathway. A very high concentration of arachidonic acid thus might still be able to result in PG synthesis.

4.5 Identification of Eicosanoid Responsible for Renin Release

The question which of the many arachidonic acid metabolites might be responsible for renin release has been studied by way of a number of different approaches. Most attention was initially directed to the possibility that PGE_2 or one of its metabolites might be responsible for PG-dependent renin release because it was a known product of renal cortical cyclooxygenase (Larsson and Anggard 1973). However, although PRA increased four fould in experiments stimulating renal release of renin by reducing blood flow to a single denervated, nonfiltering kidney, we could not discover any increase in renal venous PGE_2 by a highly specific and sensitive gas chromatographic-mass spectrometric method (Frölich 1981). This discovery prompted us to reinvestigate renal cortical arachidonate metabolism. In the course of these studies we discovered that PGI_2 is the most prominent cyclooxygenase metabolite of the renal cortex (Whorton et al. 1978). It was found to stimulate renin release in the isolated renal cortical slices at a molar concentration matched only by isoproterenol (Whorton et al. 1977). These findings led to the formulation of a hypothesis which postulates that renin release is in part mediated by PGI_2 (Frölich et al. 1978). Supportive evidence for a role of PGI_2 rather than PGE_2 as a local regulator of renin release was obtained by using an inhibitor of PGI_2 synthesis. This inhibitor caused a degree of suppression of PGI_2 synthesis comparable to its inhibitory effect on renin release following arachidonic acid exposure of the renal cortical slice, leaving isoproterenol-induced renin release intact (Whorton et al. 1980). Furthermore, infusion of PGI_2 causes renin release in dog (Gerber et al. 1978b, c, 1979; Güllner et al. 1980) and man (Fitzgerald et al. 1979; Patrono et al. 1982). This PGI_2-stimulated renin release does not seem to be due to increased blood flow, as other PGs show a more pronounced increase in RBF but less stimulation of renin (Gerber et al. 1978b). Conversely, PG synthesis inhibitors can reduce renin secretion without affecting RBF (Blackshear et al. 1979). Recent results suggest that in isolated rat glomeruli (Beierwaltes et al. 1982) and in mouse renal cortex, PGI_2 selectively stimulates the release of renin (Lin et al. 1981). Schryver et al. (1984) showed that in isolated dog glomeruli, renin release in response to arachidonic acid correlated with synthesis of 6-keto-PGF_{1a} rather than PGE_2. In addition, Jackson et al. (1982) showed a correlation between renal PGI_2 synthesis and renin secretion rate in the conscious dog subjected to acute renal artery constriction.

Other candidates for renin release originating from cyclooxygenase have been considered. PGE_2 was only weakly active in the isolated renal cortical slice (Weber et al. 1976; Whorton et al. 1981). 13, 14-dihydro-PGE_2 has also been found, on its infusion into the renal artery, to enhance renin release and was in this respect slightly more powerful than PGE_2 (Gerber et al. 1979). It is difficult to determine the true molar potencies of the various PGs because of the instability and rapid metabolism of these substances. PGI_2 tends to decompose very rapidly in aquesous solutions, while PGE_2 is quickly metabolized. PGE_1 is also a powerful stimulator of renin release in vitro (Whorton et al. 1981) and is perhaps as active as PGI_2 owing to its greater stability. However, the capacity of the kidney to produce PGE_1 appears to be limited. Renin release has been shown to be stimulated by direct infusion of PGD_2 in normal and in nonfiltering kidneys (Seymour et al. 1979). However, PGD_2 does not elicit renin from denervated, beta-adrenergic-blocked, indomethacin-treated dog kidneys (Gerber et al. 1979), despite coincident vasodilatation, and is ineffective in vitro in rabbit cortical slices.

6-keto-PGE_1 has been showen to be a stable and biologically active metabolite of PGI_2 (Wong et al. 1980). Whether it can play a role in renin release in vivo, as suggested recently (McGiff et al. 1982; Spokas et al. 1982), is an unsettled issue. It can cause renin release in vitro. However, its precursor, 6-keto-$PGF_{1\alpha}$, does not increase PRA in vivo (Patrono et al. 1982) or in vitro (Whorton et al. 1978).

5 Eicosanoids and Renal Water Metabolism

5.1 ADH Antagonized by PGE and Potentiated by Cyclooxygenase Inhibitors

The question of an interaction between ADH and PGs was first addressed in the mid-1960s, when appreciable amounts of PGE_1 became available for investigators. It had been demonstrated that PGE inhibits the effects of some hormones which stimulate lipolysis via the adenlyate cyclase/cyclic AMP (cAMP) system, and Orloff et al. (1965) extended these observations to ADH. They reported that PGE_1 inhibited the increased water permeability in response to ADH in the toad bladder. This effect of PGE_1 was also observed in the isolated rabbit collecting tubule (Grantham and Orloff 1968), and findings in the toad bladder were confirmed by other investigators. Shortly after the discovery of the PG-synthesis-inhibiting actions of nonsteroidal anti-inflammatory drugs, there came the demonstration that in the toad bladder these agents potantiate the hydro-osmotic response to ADH (Flores and Sharp 1972; Albert and Handler 1974). These results suggested that endogenous PG synthesis in the amphibian bladder normally serves to attenuate the ADH response. The hypothesis of an interaction between ADH and renal PGs was made more attractive with the demonstration that the renal medulla, which contains most of the ADH-sensitive adenlyate cyclase (Chase and Aurbach 1968), also has a very high capacity for PG synthesis (Crowshaw and Szlyk 1970; Larsson and Anggard 1973; Frölich et al. 1975a). It seemed likely, therefore, that endogenous PGs could exert effects at the primary site of ADH action, the collecting duct system; in fact, immunohistochemical techniques demonstrated the localization of PG cyclooxygenase within collecting duct cells (Janszen and Nugteren 1971; Smith and Wilkin 1977).

The first investigations designed to examine the hypothesis that endogenous PGs may act in vivo to antagonize ADH action were the studies of Anderson et al. (1975), who demonstrated that the PG synthesis inhibitors indomethacin and meclofenamate potentiate the rise in urine osmolality in response to ADH administration in the hypophysectomized dog. It was proposed that PGs may function as negative feedback inhibitors of the action of ADH. To establish the existence of this negative feedback loop at the level of the kidney, it must be demonstrated that (a) ADH produces a dose-related increase in renal PG synthesis in some portion of the physiologic range and (b) some product of PG synthesis inhibits either the release of ADH or the effect of ADH on the kidney. On the latter proposal, there is good agreement among investigators, i.e., the antagonism between ADH and PGs been consistently demonstrated through the use of a variety of preparations.

The initial approach employed by several investigators was the direct application of PGs in system where ADH's effects on water flow can be quantitated. Thus, after the early reports by Orloff et al. (1965),working with toad bladder, and Grantham and Orloff (1968), using the isolated rabbit collecting tubule, other workers demonstrated inhibitory effects of PGE_1 on ADH-stimulated water permeability (Eggena et al. 1970; Lipson and Sharp 1971; Albert and Handler 1974). As it became apparent that PGE_1 was not a major PG in mammalian kidneys, investigators examined other PGs for possible activity. Urakabe et al. (1975) compared mono- and dienoic PGs of E, F, A, and B series for their activity as inhibitors of ADH-induced water flow in toad bladder. They concluded that E and F series compounds showed much greater potency than others. Interestingly, PGE_2 and $PGF_{2\alpha}$ are the most abundant PGs in the mammalian renal medulla. Zook and Strandhoy (1980) examined their effects on both water and urea fluxes in the toad bladder. While neither PG affected basal transport, both inhibited the ADH-induced increments in water and urea movement, PGE_2 being the more potent of the two. Similar results for PGE_2 were obtained in cultured canine epithelial (MDCK) cells (Martinez and Reyes 1984) and in canine cortical collecting tubule cells (Garcia-Perez and Smith 1984).

The in vitro application of PG synthesis inhibitors has been repeatedly shown to enhance ADH-stimulated water permeability. Flores and Sharp (1972) and Parisi and Piccini (1972) reported that the non-steroidal antiinflammatory drugs aspirin and indomethacin potentiate ADH's hydrosmotic effects in the toad bladder. Actually, Nusynowitz and Forsham (1966) had previously shown a similar effect of ace-taminophen, although at that time its inhibitory effects on PG synthesis were not known. In the last 10 years, several investigators in various laboratories have shown the enhancement of ADH effects by many different nonsteroidal anti-inflammatory agents in the toad bladder (Albert and Handler 1974; Ozer and Sharp 1972; Zusman et al. 1977a; Bisordi et al. 1980; Schlondorff et al. 1981; Forrest et al. 1982). Glucocorticoids, also believed to be PG synthesis inhibitors, were shown to have similar effects (Zusman et al. 1978). Furthermore, another class of compounds not formerly regarded as PG synthesis inhibitors, the sulfonylureas, were found to have ADH-potentiating activity that seems to be mediated by PG synthesis inhibition (Zusman et al. 1977b).

The effects of indomethacin and other PG synthesis inhibitors on ADH action have also been observed in mammalian renal tissue in vitro. Ray and Morgan (1981) used the technique of collecting duct perfusion (the isolated perfused rat papilla) to study the interaction between PGs and ADH. Indomethacin shifted to the left the dose-response curve for stimulation of water permeability by ADH. Jackson et al. (1980) showed that ibuprofen and naproxen enhanced the stimulation of adenylate cyclase by ADH in isolated mammalian collecting tubules from the rat. Interestingly, this last group could not demonstrate an inhibitory effect of exogenous PGE_2 in their preparation, and it was suggested that other products of arachidonic acid oxygenation mediate the inhibitory responses revealed by nonsteroidal anti-inflammatory drugs. This question will be discussed further below.

Several studies in vivo have also examined the ADH-PG interaction. As has been previously mentioned, the direct application of PGs, either systemically or into the renal artery, is fraught with problems, especially owing to hemodynamic changes and possibly as a result of increasing

ADH release (see below). These hemodynamic changes are particularly pronounced with PGs of the E series. However, $PGF_{2\alpha}$ does not apparently share this property. Thus, Zook and Strandhoy (1981) undertook clearance studies in the anesthetized dog to assess the effects of intrarenal $PGF_{2\alpha}$ administration. Their results indicated a reduction in sodium reabsorption in Henle's loop and an inhibition of ADH-mediated water and urea absorption. The chief reservation about the importance of this effect of $PGF_{2\alpha}$ is that while hemodynamic effects may have been avoided, the distribution of the PG administered by this route may not be comparable to physiologic formation. Roman and Lechene (1981) utilized another approach by simultaneously microinjecting ^{14}C-urea and PGs into distal tubules of the meclofenamate-treated rat kidney. Urinary recovery of labeled urea was enhanced by PGE_2 or $PGF_{2\alpha}$ but not by PGA_2. Furthermore, meclofenamate treatment decreased the recovery of distally microinjected ^{14}C-urea as compared to vehicle-treated rats, thus suggesting that endogenous PGs serve to inhibit reabsorption of urea in the collecting system. The question of ADH antagonism was not specifically addressed, but since the experiments were carried out under conditions oy hypertonic saline diuresis, ADH levels were likely very high.

A more common approach to in vivo study of the ADH-PG interaction has been to assess the effects of cyclooxygenase inhibitors on water excretion. Anderson et al. (1975) first reported the potentiating effects of indomethacin or meclofenamate on ADH-induced rises in urine osmolality. Their studies were carried out in the water-loaded, hypophysectomized dogs in order to exclude the involvement of endogenous ADH. These experiments were confirmed by Fejes-Toth and coworkers (1977) in studies in water-loaded conscious dogs where inhibition of PG synthesis enhanced the antidiuretic response to exogenous ADH (lysine vasopressin). Experiments in rats yielded similar results (Lum et al. 1977; Berl et al. 1977), and the potentiation of ADH action by indomethacin or aspirin has also been observed in studies in man (Berl et al. 1977; Kramer et al. 1978; Frölich et al. 1978; Güllner et al. 1980a). In the human studies, the increment in urine osmolality above that obtained with ADH alone was generally small but very consistent.

In addition to this evidence from administration of exogenous hormone, other studies suggest that the response to endogenous ADH is also enhanced by inhibition of PG synthesis. In the studies of Fejes-Toth et al. (1977), administration of indomethacin in hydropenic conscious dogs resulted in an elevation of urine osmolality and reduction in urine flow in the absence of exogenous ADH. Similar results were later obtained by other workers in rats (Haylor and Lote 1980; Licht et al. 1981) and in man (Haylor 1980). In man, intravenous infusion of hypertonic saline was followed by an increase in urine osmolality and free water reabsorption, and indomethacin pretreatment enhanced both of these responses (Kramer et al. 1978).

A large body of evidence therefore suggests that PGs can antagonize the antidiuretic actions of ADH under a variety of experimental conditions. We will consider next the possible mechanisms of this antagonism. It should be kept in mind that some of these mechanisms may not necessarily be ADH-dependent, so that PGs may impair water reabsorption even in the absence of the hormone. Licht et al. (1981) demonstrated a reduction in urine flow by indomethacin in awake rats undergoing a water diuresis (basal urine osmolality of 89 ± 11 msmol/kg). In addition, Stoff et al. (1982) reported that indomethacin reduces free water excretion in the rat with diabetes insipidus, which is congenitally deficient in ADH (Valtin 1976). Two studies in man have demonstrated that indomethacin and aspirin impair urinary dilution during water-loading studies, reducing free water clearance by 20%–40% (Haylor 1980; Walker et al. 1981). Although the presence of small amounts of ADH cannot be rigorously excluded in these latter studies, it seems probable that under certain conditions ADH-independent mechanisms may be operative.

5.2 Mechanisms of Interaction: cAMP, Ca^{2+}, and Intramedullary Solutes

The most widely accepted mechanism by which PGs may antagonize the action of ADH is inhibition of ADH-sensitive adenylate cyclase. Such a hypothesis was based initially on the experiments of Orloff et al. (1965), which showed that PGs inhibit the response of the toad bladder to ADH but not to exogenous cAMP. A similar result was obtained in the isolated rabbit collecting tubule (Grantham and Orloff 1968). In more direct studies, PGE markely inhibited the accumulation of cAMP in response to ADH in toad bladder cells (Omachi et al. 1974), in kidney slices (Beck et al. 1971), in isolated rat papillary collecting ducts (Edwards et al. 1981) or medullary thick ascending limbs (Culpepper and Andreoli 1983; Torikai and Kurokawa 1983), and in cultured canine cortical collecting tubule cells (Garcia-Perez and Smith 1984). In addition, pretreatment with PG synthesis inhibitors enhanced the generation of cAMP in response to ADH in toad bladder (Schlondorff et al. 1981). Although the inhibition of cAMP accumulation by exogenous PGs has not been uniformly observed (Schlondorf et al. 1981; Grenier et al. 1982), the reported enhancement of cAMP levels by

PG synthesis inhibitors has been more consistent. Overall, these studies are supportive of the notion that PGs, at least endogenously formed, inhibit ADH-induced accumulation of cAMP. However, attempts at demonstrating directly the inhibition by PGs in assays of adenylate cyclase activity have not been uniformly successful. Although some early reports (Marumo and Edelman 1971; Beck et al. 1971; Kalisker and Dyer 1972) indicated such an inhibition by PGs, the magnitude of the response was very small, and other investigators failed to demonstrate any activity (Birnbaumer and Yang 1974; Herman et al. 1979; Jackson et al. 1980; Morel et al. 1980; Torikai and Kurokowa 1981). In fact, many reports indicate that PGs can stimulate renal medullary adenylate cyclase, at least at high concentrations (Birnbaumer and Yang 1974; Zenser and Davis 1977; Herman et al. 1979; Torikai and Kurokawa 1981; Edwards et al. 1981). This failure to observe the inhibitory effects of PGs on ADH-sensitive adenylate cyclase may be due to assay conditions, as the adenylate cyclase activity measurements are made in broken cell preparations where the integrity of an inhibitory complex may be disrupted. Alternatively, the abovementioned stimulation of adenylate cyclases by PGs may obscure any inhibitory effects that may be restricted to a certain cell type or a certain receptor-enzyme type.

Another explanation is suggested by some recent experiments. Herman et al. (1980) reported that PGH_2, one of the endoperoxide precursors of PGE_2, markedly inhibited the stimulation of renal medullary adenylate cyclase by ADH. The inhibition was short-lived, as is consistent with the rapid metabolism of PGH_2-stimulated adenylate cyclase activity, and was additive with ADH. The authors concluded that PGH_2 might act in vivo to modulate ADH action. Although the high concentrations (10^{-4} M) of PGH_2 used raise concern about the physiologic significance of this response, other studies are consistent with the proposal that PGH_2 (or PGG_2) may be an inhibitors of ADH-sensitive adenylate cyclase. As previously mentioned, Jackson et al. (1980) could not demonstrate inhibitory effects of exogenous PGE_2 or PGF_{2a} on ADH-stimulated adenylate cyclase activity in membranes of isolated rat medullary collecting tubules. However, if the preparations were preincubated with ibuprofen or naproxen, the stimulatory effect of ADH on adenylate cyclase was markedly potentiated. These workers also suggested that the endoperoxides might be logical candidates for the endogenous inhibitors. Gorman (1975) reported on the inhibition of the adenylate cyclase of adiopocyte ghosts by PGH_2 but not PGE_2, a finding similar to that of Herman et al. (1980). Thus, although a role for the endoperoxides is speculative, further studies seem warranted.

It might also be noted that the effects of intracellularly formed endo-peroxides (or other PGs) may be substantially different from those obtained by extracellular addition as is done in most in vivo studies. Ray and Morgan (1981) observed that in the isolated rat papilla, addition of exogenous PGE_2 increased water permeability, consistent with the stimulation of a PG-sensitive adenylate cyclase. However, treatment with indomethacin caused a shift to the left of the ADH-water permeability concentration-response curve, so that endogenous PGs appear to have an opposite effect to that of exogenously added PGs, at least under these conditions.

Antagonism of ADH responses at a step distal or unrelated to generation has also been suggested as a mechanism. Schlondorff et al. (1981) reported that in the toad bladder PGE_2 increases cAMP content and activates protein kinase to an extent similar to that achieved by ADH, but does not enhance water flow. Their suggestion was that PGE_2 interferes with the action of protein kinase to alter water permeability. There is little other evidence to support this notion, but the question has not yet been extensively pursued. One possible cAMP-independent mechanism by which PGs may inhibit water permeability responses to ADH is impairment of calcium transport. As some recent evidence suggests a critical role for intracellular calcium (Humes et al. 1980; Hardy 1978) and calcium calmodulin (Levine et al. 1981; Grosso et al. 1982) in the action of ADH, this pathway represents a potential site of PG-ADH interaction. The role of calcium, however, in ADH action is incompletely understood. It appears that extracellular calcium is not required for its effects, except to replenish cellular stores (Burch and Halushka 1983b). ADH did reduce the size of a slowly exchanging kinetic pool of calcium in tod bladder epithelial cells (Burch and Halushka 1983a). Interestingly, agents which mimic the hydroosmotic effect of ADH decreased the size of this same pool, while PGE increased it. It appears that PGs may modulate ADH action by altering intracellular calcium distribution, but this interaction requires more study.

Another possible mechanism by which PGs may inhibit the reabsorption of water is via alterations in medullary solute accumulation. As the reabsorption of water by the collecting ducts is passive (down its concentration gradient), any manipulation which decreases that gradient would impair the water reabsorptive response. At least three separate renal actions of PGs might account for a reduced solute content in the renal medulla (see Stokes 1981). The first is the action of PGs to increase medullary blood flow, as alluded to in Sect. 2. Although the effects of PGs on RBF are not completely resolved, several studies have indicated that indomethacin or other PG synthesis inhibitors decrease medullary flow under some conditions (Solez et al. 1974; Ganguli et al. 1977; Düsing et al. 1977; Chuang et al. 1978). In addition, many other studies have demonstrated that indomethacin may decrease blood flow in inner cortical nephrons (see Sect. 2), and insofar as the inner cortical circulation reflects medullary blood flow, they suggest that PGs may serve to maintain medullary perfusion. Any increase in medullary blood flow would serve to "wash out" the solute content of the papillary interstitium, as the low flow rates in the papilla are essential to the accumulation of the osmotic gradient.

Reduction of tubular sodium chloride reabsorption is another mechanism by which PGs may impair the maintenance of medullary solute content. Currently available evidence on this point has been reviewed in Sect. 3. Of special interest for this discussion are the findings that PGE_2 inhibits sodium and chloride transport in isolated medullary, but not cortical, thick ascending limbs (Stokes 1979; Culpepper and Andreoli 1983) and in isolated medullary collecting tubules (Stokes and Kokko 1977; Iino and Imai 1978). It is interesting that Culpepper and Andreoli (1983) found that this effect of PGE_2 was only observed in the presence of ADH, which stimulates sodium chloride reabsorption in this tubular segment. The half-maximal inhibitory concentration occurred at 10^{-10} M PGE_2, thus suggesting a potential physiologic importance. These effects, if operative in vivo, would directly impair sodium reabsorption at the sites of the active transport step in the operation of the countercurrent mechanism and could therefore lead to a reduced osmotic driving force for water reabsorption. Some in vivo studies are consistent with this proposal, since PG synthesis inhibitors enhance accumulation of sodium and chloride in the renal medulla (Ganguli et al. 1977) and increase Henle's loop sodium reabsorption during volume expansion in the rat (Higashihara et al. 1979).

Also critically involved in the operation of the countercurrent system is urea. As it is the primary solute of the inner medullary interstitium, an effect on urea deposition could have pronounced effects on collecting duct water reabsorption. According to current concepts, urea permeability in distal nephron segments is fairly low during antidiuresis, except in the papillary collecting ducts (Rocha and Kokko 1974). Urea is therefore concentrated by abstraction of water along the distal nephron until it reaches the papilla, where the higher permeability allows it to diffuse passively down its concentration gradient and deposit in the papillary interstitium. Several studies suggest that PGs may interfere with this reabsorption of urea. Carvounis et al. (1979) and Zook and Strandhoy (1980) reported that in the toad bladder, PGE_2 and $PGF_{2\alpha}$ inhibit the increase in urea flux induced by ADH.

Zook and Strandhoy (1981) also reported that intrarenal infusion of $PGF_{2\alpha}$ in hydropenic dogs causes depletion of the corticopapillary urea gradient in the absence of detectable intrarenal hemodynamic changes. Roman and Lechene (1981), utilizing micropuncture techniques in the rat, demonstrated an enhanced reabsorption of distally micro-injected ^{14}C-urea following PG synthesis inhibition. Addition of PGE_2 or $PGF_{2\alpha}$ to the injectate reversed the increment in reabsorption, in-dicating that urea deposition into the renal papilla may be directly diminished by these PGs. Furthermore, increased medullary and papil-lary urea concentrations have been observed after indomethacin ad-ministration (Haylor and Lote 1983). Which of these mechanisms may be operative in the antagonism by PGs of ADH-induced water reabsorp-tion is unknown. Stokes (1981) and Roman and Lechene (1981) pointed out that in situations where ADH levels are maximal, virtually no barrier to water reabsorption exists in the collecting duct and there is complete osmotic equilibration with the medullary interstitium. In such condi-tions, inhibition of PG synthesis is still effective in enhancing antidiuresis, thus suggesting that mechanisms other than changes in collecting duct water permeability (i.e., alteration of medullary osmotic gradients) must be involved. However, it would seem that at lower ADH con-centrations, enhanced water permeability has not been excluded.

These experimental observations which showed potentiation of the hydro-osmotic actions of ADH by inhibitors of PG synthesis have led to the use of these drugs in some clinical states where hyporesponsiveness to ADH is apparent. Thus several investigators have demonstrated beneficial effects of indomethacin or other PG synthetase inhibitors in patients with nephrogenic diabetes insipidus, whether hereditary (Blachar et al. 1980; Usverti et al. 1980; Fichman et al. 1977) or lithium-induced (Frölich et al. 1977; Walker et al. 1980). One pathologic con-dition which has been widely investigated is the urinary concentration defect associated with hypokalemia. Results indicate that inhibition of PG synthesis does not improve the maximal response to ADH in hypo-kalemic patients, including those with Bartter's syndrome, diuretic-induced hypokalemia, "cryptogenic" renal potassium loss, primary aldosteronism, and laxative abuse (Rodriguez et al. 1981), or in hypo-kalemic dogs (Rutecki et al. 1982). These reports are in good agreement with previous findings in the rat (Hood and Dunn 1978; Berl et al. 1980) that the concentrating defect in rats on a low potassium diet is PG-independent. It should be emphasized that in many of these studies, PG synthesis inhibition decreased urine flow, but maximal urinary concentration was not affected. Therefore, although PGs may con-tribute in some fashion to the increased urine flow, they do not appear to account for the decreased sensitivity of the collecting duct to ADH. In studies in the dog made hypokalemic by dialysis (Galvez et al. 1977) or by administration of desoxycorticosterone acetate (Düsing et al. 1982), indomethacin also decreased urine flow, but the maximal concentration of urine in response to ADH was not tested.

5.3 PGE$_2$ as Negative Feedback Inhibitor of ADH

According to the proposed scheme for negative feedback inhibition by PGs of ADH action, increased circulating levels of ADH − or some consequence thereof − would enhance PG synthesis. Many studies have addressed this question, but much controversy remains. The earliest studies supportive of the concept were carried out in vivo in medullary slices (Kalisker and Dyer 1972), where addition of ADH at high concentrations stimulated PGE release. Zusman and Keiser (1977b) reported similar findings in cultured rabbit renomedullary interstitial cells, but again fairly high concentrations were used (10^{-7} M). Zusman et al. (1977a) demonstrated that in the toad bladder PGE biosynthesis was stimulated by ADH in more physiologic concentrations, well below the level required for half-maximal stimulation water flow. Burch et al. (1979) confirmed the stimulation of PG production by ADH in toad bladder. However, utilizing similar preparations, Bisordi et al. (1980) and Forrest et al. (1982) were unable to demonstrate stimulation of PG biosynthesis by ADH. The reason for these discrepancies is not presently clear.

More recently, Burch and Halushka (1982) showed that ADH increases PGE and TXB$_2$ biosynthesis in toad bladder epithelial cells. Several mammalian systems have also been studied in vitro. ADH stimulated PG biosynthesis in cultured rat renomedullary interstitial cells (Beck et al. 1980) and glomerular epithelial cells (Lieberthal and Levine 1984) in the rabbit kidney perfused ex vivo (Zipser et al. 1981b). Grenier et al. (1981) did not observe any effect of ADH on PG synthesis in cultures of rabbit papillary collecting tubule cells, but the same group (Garcia-Perez and Smith 1984) obtained positive results in canine cortical collecting tubule cells in culture.

In vivo studies have been more difficult to interpret. Experiments in the rat have generally given consistent support for the stimulation of PG synthesis by ADH. Of particular interest are studies in the rat with hereditary diabetes insipidus. It has been demonstrated that in these animals, which lack endogenous ADH, urinary PG excretion is lower than in normal rats (Walker et al. 1978; Dunn et al. 1978a).

Furthermore, chronic administration of ADH or dDAVP, the nonpressor analog of ADH, results in restoration of PG excretion to normal levels (Walker et al. 1978a; Dunn et al. 1978; Walker et al. 1981; Beck et al. 1982). The enhancement of PGE_2 excretion by ADH treatment was also observed in normal rats (Walker et al. 1978).

In patients with central diabetes insipidus, an increase in PGE excretion with dDAVP administration was observed (Düsing et al. 1980). Also, the same group (Kramer et al. 1978) observed an increase in PGE_2 excretion with lysine vasopressin infusion in patients previously given hypertonic saline to suppress PRA. They proposed (Kramer et al. 1981) that in man, angiotensin II is a more potent stimulator of renal PG synthesis than is ADH and that varying activities of the renin-angiotensin system may thus obscure ADH responses if not carefully controlled. Other studies carried out in man and in the dog did not provide consistent evidence that ADH stimulates renal PG synthesis. Thus Walker et al. (1981) and Kirschenbaum and Serros (1980) did not observe increased urinary PGE excretion with ADH administration to water-loaded subjects. However, these authors and others (Kaye et al. 1980; Wright et al. 1981; Fejes-Toth et al. 1983a) reported a positive correlation between urine flow and PGE excretion in dog and man. In a more recent study, both AVP and dDAVP were found to stimulate urinary PG excretion in dogs during constant urine flow rate (Fejes-Toth et al. 1983a). This and previous findings from the same laboratory (Fejes-Toth et al. 1977) indicate that ADH can increase renal PG synthesis in dogs, but in water diuresis the decrease in PG excretion as a result of the reduced urine volume obscures the effect of ADH.

The question of stimulation of PG biosynthesis by ADH is further complicated by consideration of pressor activity of arginine vasopressin. It is well known that ADH is a powerful vasoconstrictor, at least at high concentrations. Since other vascoconstrictors (angiotensin II, no-repinephrine) are known to stimulate PG synthesis in various vascular beds, it has been considered that ADH may have a similar effect. This indeed appears to be the case, as the effect of ADH on PG production in several in vitro systems is related to its pressor, rather than its anti-diuretic, action. Thus, in rat renomedullary interstitial cells (Beck et al. 1980), glomerular epithelial cells (Lieberthal and Levine 1984, and mesangial cells (Uglesity et al. 1983), as in the ex vivo perfused rabbit kidney (Zipser et al. 1981b), ADH increased PG release, but the non-pressor analog dDAVP was without effect. Some in vivo studies have also lent support to this concept. Zipser et al. (1981a) examined urinary PGE_2 excretion in normal human volunteers during ADH or dDAVP administration. They demonstrated an increase in PG excretion with ADH but a decrease with an equal (in terms of antidiuretic activity) dose of dDAVP. As angiotensin infusion also increased urinary PG excretion in their protocol, they concluded that the stimulation of PG synthesis by ADH was accounted for by its pressor action. Nadler et al. (1983) came to similar conclusions in studies in healthy women. These findings led to the proposal (Beck et al. 1982; Bankir et al. 1980) that dDAVP only increases renal PG synthesis via induction of cyclo-oxygenase and lacks an acute stimulatory effect on arachidonate availability for PG synthesis. However, evidence is accumulating that this is not the case. dDAVP does have an acute stimulatory effect on PG excretion in hydropenic dogs (Fejes-Toth et al. 1983a) and in rats with hereditary diabetes insipidus or in normal rats (Fejes-Toth et al. 1983b). In these studies, PGE_2 excretion increased within minutes of ADH or dDAVP administration. These studies indicate that neither vasoconstriction nor enzyme induction, which would be require more time, are necessary for the increase in urinary PGE_2. Furthermore, Burch and Halushka (1982) reportet that dDAVP acutely increases PG biosynthesis in toad bladder epithelial cells, and Kirschenbaum et al. (1982) demonstrated a similar result in the isolated rabbit cortical collecting tubule. The explanation for the lack of dDAVP activity in many systems is not clear. It is possible that the vascular and interstitial elements of the kidney respond to the pressor activity of ADH in a fashion similar to that of other vasoactive hormones (angiotensin, bradykinin) and that in tubular elements, PG synthesis is activated by the antidiuretic receptor.

In conclusion, the importance of PGs in modulating the renal response to ADH seems well established. Studies in more well defined cell populations and determination of the exact origin of urinary PGs under various conditions in several species should provide a better understanding of the interaction.

6 Renal Prostanoids and Erythropoietin Production

PGE_2 can stimulate erythropoietin production in the intact animal (Gross et al. 1976) and the isolated perfused kidney (Schooley and Mahlmann 1971). More recently, the mesangial cell has been shown to be a source of erythropoietin and responsive to PGE_2, PGI_2, arachidonic acid, and hypoxic stimulation with respect to production of the hormone (Kurtz et al. 1983; Kurtz et al. 1985). It was demonstrated that low O_2 concentrations stimulate PGE_2 synthesis in these cells as well as erythropoietin production, and that both effects were abolished by inhibition of cyclooxygenase by indomethacin (Kurtz et al. 1985). The indomethacin effect on erythropoietin production could be over-come by addition of exogenous PGE_2, an observation which nicely demonstrates that it was the effect of indomethacin on cyclooxygenase rather than one of its other effects which was responsible for suppression of hypoxic erythropoietin stimulation (Kurtz et al. 1985). A mechanism involving stimulation of cAMP for the enhanced erythropoietin produc-tion of these cells has been suggested, since PGE_2, and arachidonic acid stimulate cAMP, and forskolin, which directly stimulates the catalytic subunit of adenylate cyclase, also stimulates erythropoietin production in this model (Kurtz et al. 1985).

7 Leukotrienes

Leukotrienes (LTs) are a family of compounds derived from arachidonic acid which have been shown to represent the major part of the slow-reacting substance of anaphylaxis (for review see Samuelsson 1982). Since their identification, numerous biological activities have been described, including bronchoconstriction, increase in vascular permea-bility, vasoconstriction, chemotaxis, chemokinesis, adhesion, enzyme release, and superoxide generation by granulocytes, which make LTs prime candidates as mediators of inflammation and hypersensitivity reactions (for review see Samuelsson 1983; Lewis and Austen 1984) as well as shock reactions (Lefer 1986). Their biosynthesis and metab-olism have been studied (for review see Hammarström 1983).

In the recent past, considerable attention has been directed to the possibility that LTs affect renal function. This interest stems in part from observations on single-dose intravenous injections of LTC_4, which cause a dose-dependent increase in urine flow, sodium and potassium excretion, and changes in renal hemodynamics in the unanesthetized rat (Filep et al. 1984; Filep et al. 1985). Blood pressure increased, while RBF decreased and GFR showed an increase at the highest con-centration of LTC_4. Others giving LTC_4 to anesthetized rats at a con-stant rate of infusion noticed an increase in blood pressure, a decrease of 20% in blood volume and a decrease in GFR. The decrease in GFR was interpreted as a consequence of the reduced blood volume and increased angiotensin II, as it was reversed by the angiotensin II-receptor antagonist, saralasine (Badr et al. 1984). While it was possible to antagonize the vasoconstriction and renal excretory effects by pretreatment with the LT antagonist FPL 55712, this drug by itself had no effect on any of the parameters under observation (Filep et al. 1985). This strongly suggests that renal cysteinyl LTs are not released under basal conditions in quantities sufficient significantly to affect these parameters. This is in striking contrast to renal PGs, which play a role in a variety of physiologic functions even under basal conditions.

Recent evidence indicates that LT responses are mediated by specific receptors, which have been identified in rat renal glomeruli (Ballermann et al. 1985). The existence of such receptors was made likely by the renal responses to LTC_4 and their abolition by the receptor antagonist FPL 55712 (Filep et al. 1985). Furthermore, the presence of these receptors raises the possbility that they might be stimulated by endogenous LT synthesis.

While renal vasoconstriction has been observed in the rat (Rosenthal and Pace-Asciak 1983), renal vasodilation has been seen after LTC_4 or LTD_4 administration in the dog (Feigen 1983; Chapnick 1984; Secrest et al. 1985). This vasodilator reponse to LTD_4 was endothelium-dependent (Secrest et al. 1985). Thus, while vascular LT responses are frequently coupled to prostanoid responses, for example, in the rat kidney, LTC_4 vasoconstriction can be blocked by a TXA_2 antagonist (Ogletree et al. 1985), this is the first report of a coupling of LTC_4 response to the non-cyclooxygenase endothelium-derived relaxing factor (EDRF) (Secrest et al. 1985).

Lipoxygenase products have been identified as 12-hydroxy-eicosatetraenoic acid (12-HETE, 11-HETE, 15-HETE, and 8- or 9-HETE) in isolated glomeruli, glomerular epithelial cells, and cortical tubules synthesized in normal kidneys from endogenous substrate (Winokur and Morrison 1981; Jim et al. 1982; Sraer et al. 1983). It is presently not known whether these substances are synthesized in vivo. Synthesis of LTC_4 has recently been described in renal allograft rejection (Coffmann et al. 1986), and synthesis of LTB_4 and LTC_4 in nephrotoxic serum glomerulonephritis (Lianos 1986). While there are present in the latter biological models a large number of granulocytes and macrophages, providing a possible source of these LTs, Sraer et al. (1986) showed that even normal glomeruli are able to convert arachidonic acid to LTC_4 in vitro. These findings raise the possibility that LTs play a major role in renal inflammatory disease and in responses to hemorrhage and other noxious stimuli. A major question will be the differentiaton of LT synthesis by local resident cells as compoared to blood-borne cells. Renal LTs are now reaching the stage of concerted investigative efforts. They, together with other recently identified local mediators such as platelet activating factor and EDRF, promise to unlock some of the many secrets of renal pathophysiology.

Acknowledgements. We wish to thank Dr. J. Filep and Dr. J. Fauler for valuable criticism. Mrs. I. Koch, Pam Shoemaker, and Paula Norris solved the secretarial nightmare brilliantly.

References

Anggard E, Larsson C (1971) The sequence of the early steps in the metabolism of prostaglandin E. Eur J Pharmacol 14:66–70

Anggard E, Bohman SO, Griffin JE, Larsson C, Maunsbach AB (1972) Subcellular localization of the prostaglandin system in the rabbit renal papilla. Acta Physiol Scand 84:231–246

Aiken JW, Vane JR (1973) Intrerenal prostaglandin release attenuates the renal vasoconstrictor activity of angiotensin. J Pharmacol Ther 184:678–687

Albert W.C., Handler JS (1974) Effects of PGE_1, indomethacin and polyphloretin phosphate on toad bladder response to ADH. Am J Physiol 226:1382–1386

Allen JT, Vaughn ED Jr, Gillenwater JY (1978) The effect of indomethacin on renal blood flow and ureteral pressure in unilateral ureteral obstruction in awake dogs. Invest Urol 15:324–327

Altsheler P, Klahr S, Rosenbaum R, Slatopolsky E (1978) Effects of inhibitors of prostaglandin synthesis on renal sodium excretion in normal dogs and dogs with decreased renal mass. Am J Physiol 235:338–344

Anderson RJ, Berl T, McDonald KM, Schrier RW (1975a) Evidence for an in vivo antagonism between vasopressin and prostaglandin in the mammalian kidney. J Clin Invest 56:420–426

Anderson RJ, Taber MS, Cronin RE, McDonald KM, Schrier RW (1975) Effect of B-adrenergic blockade and inhibitors of angiotensin II and prostaglandin on renal autoregulation. Am J Physiol 229:731–736

Arisz L, Donker AJM, Brentjens JRH, van der Hem GK (1976) The effect of indomethacin on proteinuria and kidney function in the nephrotic syndrome. Acta Med Scand 199:121–125

Arroyo V, Planas R, Gaya J, Deulofew R, Rimola A, Perez-Ayuso R, Rivera F, Rodes J (1983) Sympathetic nervous activity, renin-angiotensin system, and renal excretion of prostaglandins E_2 in cirrhosis: relationship to functional renal failure and sodium and water excretion. Eur J Clin Invest 13:271–278

Badr KF, Baylis C, Pfeffer JM, Pfeffer MA, Soberman RJ, Lewis RA, Austen KF, Corey EJ, Brenner BM (1984) Renal and systemic hemodynamic responses to intravenous infusion of leukotriene C_4 in the rat. Circ Res 54:492–499

Baer PG, McGiff JC (1979) Comparison of eccts of prostaglandin E_2 and I_2 on rat renal vascular resistance. Eur J Pharmacol 54:359–363

Bailie MD, Barbour JA, Hook JB (1975) Effects of indomethacin on furosemide-induced changes in renal blood flow. Proc Soc Exp Biol Med 148:1173–1176

Ballermann BJ, Lewi RA, Corey EJ, Austen KF, Brenner BM (1985) Identification and characterization of LTC_4 receptors in isolated rat glomeruli. Circ Res 56:324–330

Bankir LR, Trinh-Trang-Tan MM, Nivez MP, Sraer J, Ardaillou R (1980) Altered PGE_2 production by glomeruli and papilla of rats with hereditary diabetes insipidus. Prostaglandins 20:349–365

Banks RA, Beilin LH, Soltys JS (1980) The effects of meclofenamate on renal and peripheral vascular beds in conscious rabbits. Uin Exp Pharmacol Physiol 7:531–534

Barchowsky A, Data JL, Whorton AR (1984) The effect of prostaglandin synthesis inhibition on the direct stimulation of renin release from rabbit renal cortical slices. Prostaglandins 27:51–67

Bartelheimer HK, Senft G (1968) Zur Lokalisation der turbulären Wirkung einiger antirheumatisch wirkender Substanzen. Arzneimittelforschung 18:567−570

Bartels J, Kunz H, Vogt W, Wille G (1970) Prostaglandin: liberation from and formation in perfused frog intestine. Naunyn-Schmiedebergs Arch Exp Pathol Pharmakol 266:199−209

Battez G, Goulet L (1913) Action des l-extrait de prostate humaine sur la vessie et sur la pression arterielle. Compt Rend Soc Biol 74:8−9

Bayliss C, Brenner BMK (1978) Modulation by prostaglandin synthesis inhibitors of the action of exogenous angiotensin II on glomerular ultrafiltration in the rat. Circ Res 43:889−898

Beck NP, Kaneko T, Zor U, Field JB, Davis BB (1971) Effect of vasopressin and prostaglandin E, on the adenyl cyclase-cyclic $3'-5'$-adenosine monophosphate system of the renal medulla of the rat. J Clin Invest 50:2461−2465

Beck TR, Hassid A, Dunn MJ (1980) The effect of arginine vasopressin and its analogues on the synthesis of PGE_2 by rat renal medullary interstitial cells in culture. J Pharmacol Exp Ther 215:15−19

Beck N, Shaw JO (1981) Thromboxane B_2 and prostaglandin E_2 in the K^+-depleted rat kidney. Am J Physiol 240:151−157

Beck TR, Hassid A, Dunn MJ (1982) Desamino-D arginine vasopressin induces fatty acid cyclooxygenase activity in the renal medulla of diabetes insipidus rats. J Pharmacol Exp Ther 221:269−274

Beierwaltes WH, Schryver S, Olson PS, Romero JC (1980) Interaction of the prostaglandin and renin-angiotensin systems in isolated rat glomeruli. Am J Physiol 239:F602−F608

Beierwaltes WH, Schryver S, Sanders E, Strand J, Romero JC (1982) Renin relase selectively stimulated by prostaglandin I_2 in isolated rat glomeruli. Am J Physiol 243:F276−F283

Beilin JL, Bhattacharya J (1977a) The effect of prostaglandin synthesis inhibitors on renal blood flow distribution in conscious rabbits. J Physiol 269:293−405

Beilin LJ, Bhattacharya J (1977b) The effect of idomethacin on auto-regulation of renal blood flow in the anesthetized dog. J Physiol 271:625−639

Bell C, Mya MKK (1977) Release by vasopressin of E-type prostaglandins from the rat kidney. Clin Sci Mol Med 52:103−106

Bell RL, Kennerly DA, Stanford N, Majerus PW (1979) Diglyceride lipase: a pathway for arachidonate release from human platelets. Proc Natl. Acad Sci USA 76:3238−3241

Benabe JE, Klahr S, Hoffmann MK, Morrison AR (1980) Production of thromboxane A_2 by the kidney in glycerol-induced acute renal failure. Prostaglandins 19:333−347

Benabe JE, Spry LA, Morrison AR (1982) Effects of angiotensin II on phosphatinositol and polyphosphoinositide turnover in rat kidney: mechanism of prostaglandin release. J Biol Chem 257:7430−7434

Berg KJ (1977a) Acute effects of acetylsalicylic acid in patients with chronic renal insufficiency. Eur J Clin Pharmacol 11:111−116

Berg KJ (1977b) Acute effects of acetylsalicylic acid on renal function in normal man. Eur J Clin Pharmacol 11:117−123

Bergström S, Sjövall J (1969) The isolation of prostaglandin E from sheep prostate glands. Acta Chem Scand 14:1071−1705

Bergström S, Dressler F, Krabisch L, Ryhage R, Sjövall J (1962a) The isolation and structure of a smooth muscle stimulating factor in normal sheep and pig lungs. Ark Kemi 20:63−66

Bergström S, Dressler F, Ryhage R, Samuelsson B, Sjövall J (1962b) The isolation of two further prostaglandins from sheep prostate glands. Ark Kemi 19:563−567

Bergström S, Ryhage R, Samuelsson B, Sjövall J (1962c) The structure or prostaglandin E, F_1 and F_2. Acta Chem Scand 16(2):501−528

Bergström S, Danielsson H, Samuelsson B (1963a) The enzymatic formation of prostaglandin E_2 from arachidonic acid: Prostaglandins and related factors 32. Biochem Biophys Acta 90:207–210

Bergström S, Rhygage R, Samuelsson B, Sjövall J (1963b) Prostaglandins and related factors: the structures of prostaglandin E_1, F_{1a} and F_1. J Biol Chem 238: 3555–3564

Berl T, (1981) Cellular calcium uptake in the action of prostaglandin on renal water excretion. Kidney Int 19:15–23

Berl T, Schrier RW (1973) Mechanism of effect of prostaglandin E_1 on renal water excretion. J Clin Invest 52:463–471

Berl T, Raz M, Wald H, Horowitz J, Cazckes W (1977) Prostaglandinsynthesis inhibition and the action of vasopressin: studies in man and rat. Am J Physiol 232:529–537

Berl T, Henrich WL, Erickson AL, Schrier RW (1979) Prostaglandins in the beta-adrenergic and baroreceptor-mediated secretion of renin. Am J Physiol 236: F472–F477

Berl T, Erickson AL, Schrier RW (1980) In vivo evidence for a role of cellular calcium (Ca) in the hydroosmotic response to vasopressin (VP). Clin Res 28:437A

Berl T, Aisenbray GA, Linas SL (1980) Renal concentrating defect in the hypokalemic rat is prostaglandin independent. Am J Physiol 238:F37–F41

Billah MM, Lapetina EG, Cuatrecasas P (1980) Phospholipase A_2 and phospholipase C activities of platelets: differential substrate specificity, Ca^{2+} requirement, pH dependence, and cellular localization. J Biol Chem 255:10227–10231

Bills TK, Smith JB, Silver MJ (1976) Metabolism of arachidonic acid by human platelets. Biochim Biophys Acta 424:303–314

Bills TK, Smith JB, Silver MJ (1977) Selective release of arachidonic acid from the phospholipids of human platelets in response to thrombin. J Clin Invest 60:1–6

Birnbaumer L, Yang PC (1974) Studies on receptor-mediated activation of adenylyl cyclase. J Biol Chem 249:7848–7856

Bisordi JE, Schlondorff D, Hays RM (1980) Interaction of vasopressin and prostaglandins in the toad urinary bladder. J Clin Invest 66:1200–1210

Blachar Y, Zadik Z, Shemesh M, Kaplan BS, Levin S (1980) The effect of inhibition of prostaglandin synthesis on free water and osmolar clearances in patients with hereditary nephrogenic diabetes insipidus. Int J Pediadr Nephrol 1:48–51

Blackshear JL, Spielman WS, Knox FG, Romero JC (1973) Dissociation of renin release and renal vasodilation by prostaglandin synthesis inhibitors. Am J Physiol 6:F20–F24

Blaine EH, Davis JO, Witty RT (1970) Renin release after hemorrhage and after suprarenal aortic constriction in dogs without sodium delivery to the macula densa. Circ Res 27:1081–1089

Blasingham MC, Nasjletti A (1980) Differential renal effects of cyclo-oxygenase inhibition in sodium-replete and sodium-deprived dog. Am J Physiol 239: 360–365

Blasingham MC, Shade RE, Share L, Nasjletti A (1980) The effect of meclofenamate on renal blood flow in the unanesthetized dog: relation to renal prostaglandins and sodium balance. J Pharmacol Exp Ther 214:1–4

Bolger PM, Eisner GM, Ramwell PW, Slotkoff LM (1976) Effect of prostaglandins synthesis on renal function and renin release in the dog. Nature: 259:244–245

Bolger PM, Eisner GM, Shea PT, Ramwell PW, Slotkoff LM (1977) Effects of prostaglandin D_2 on canine renal function. Nature 267:628–630

Bolger PM, Eisner GM, Ramwell PW, Slotkoff LM, Corey EJ (1978) Renal actions of prostacyclin. Nature 271:467–469

Bowden RE, Gill JR Jr, Radfar N, Taylor AA, Keiser HR (1978) Prostaglandin synthetase inhibitors in Bartter's syndrome. Effect on immunoreactive prostaglandin E excretion. JAMA 239:117–121

Boyer TD, Zia P, Reynolds TB (1979) Effect of indomethacin and prostaglandin A_1 on renal function and plasma renin activity in alcoholic liver disease. Gastroenterology 77:215–222

Brash AR, Jackson ER, Saggese C, Lawson JA, Oates JA, Fitzgerald GA (1983) The metabolic disposition of prostacyclin in man. J Pharmacol Exp Ther 226:78–87

Brater DC (1973) Effect of indomethacin on salt and water homeostasis. Clin Pharmacol Exp Ther 25:322–330

Brater DC, Beck NM, Adams BV, Campbell WB (1980) Effects of indomethacin on furosemide-stimulated urinary PGE_2 excretion in man. Eur J Pharmacol 65: 213–219

Briggs JP, Wright FS (1979) Feedback control of glomerular filtration rate: site of the effector mechanism. Am J Physiol 236:F40–F47

Bundy GL, Daniels EL, Lincoln FH, Pike JE (1972) Isolation of a new naturally occurring prostaglandin, 5-trans-PGA_2. Synthesis of 5-trans-PGE_2 and 5-trans-PGF_2. J Am Chem Soc 94:2124

Burch RM, Knapp DR, Halushka PV (1979) Vasopressin stimulates thromboxane synthesis in the toad urinary bladder: effects of imidazole. J Pharmacol Exp Ther 210:344–348

Burch RM, Halushka PV (1982) Vasopressin stimulates prostaglandin and thromboxane synthesis in toad bladder epithelial cells. Am J Physiol 243:593–597

Burch RM, Halushka PV (1983a) ^{45}Ca fluxes in isolated toad bladder epithelial cells: effects of agents which alter water or sodium transport. J Pharmacol Exp Ther 224:108–117

Burch RM, Halushka PV (1983b) Verapamil inhibition of vasopressinstimulated water flow: possible role of intracellular calcium. J Pharmacol Exp Ther 226: 701–705

Cadnapaphornchai P, Schrier RW, Aisenbrey G, McDonald KM, Burke TJ (1978) Prostaglandins-mediated hyperemia and renin-mediated hypertension during acute ureteral obstruction. Prostaglandins 16:965–971

Campbell WB, Graham RM, Hackson EK (1979) Role of renal prostaglandins in sympathetically mediated renin release in the rat. J Clin Invest 64:448–456

Carriere S, Thorburn GD, O'Morchoe CCC, Barger CA (1966) Intrerenal distribution of blood flow in dogs during hemorrhagic hypotension. Circ Res 19:167–179

Caruna RJ, Hudson JB, Fowler BL (1980) The effect of indomethacin (I) or urinary calcium excretion ($U_{Ca}V$) in normal adults. Clin Res 28:440A

Carvounis CP, Franki N, Levine SD, Hays RM (1979) Membrane pathways for water and solutes in the toad bladder: independent activation of water and urea transport. J Membr Biol 49:253–268

Catterton Z, Sellers B Jr, Gray B (1980) Inulin clearance in the premature infant receiving indomethacin. J Pediatr 96:737–739

Chang LCT, Splawinsky JA, Oates JA, Nies AS (1975) Enhanced renal prostaglandin production in the dog II: Effects on intrarenal hemodynamics. Circ Res 36: 204–207

Chapnick BM (1984) Divergent influences of leukotrienes C_4, D_4 and E_4 on mesenteric and renal blood flow. Am J Physiol 246:H518–H524

Chase LR, Aurbach GD (1968) Renal adenyl cyclase: anatomically separate sites for parathyroid hormone and vasopressin. Science 159:545–547

Chuang EL, Rineck HJ, Osgood RW, Kunau RT, Stein JH (1978) Studies on the mechanism of reduced urinary osmolality after exposure of the renal papilla. J Clin Invest 61:633–639

Ciabattoni G, Ginotti GA, Pierucci A, Simonetti BM, Manzi M, Pugliese F, Barsotti P, Pecci G, Taggi F, Patrono C (1984) Effects of sulindac and ibuprofen in patients with chronic glomerular disease. N Engl J Med 310:279–83

Cifuentes RF, Olley PM, Balfe JW, Radoe IC, Soldin SJ (1975) Indomethacin and renal function in premature infants with non-steroidal anti-inflammatory drugs. N Engl J Med 310:563–572

Clive DM, Stoff JS (1984) Renal syndrome associated with non-steroidal anti-inflammatory drugs. N Engl J Med 310:563–572

Coffman TM, Yarger WE, Kotman PE (1986) Leukotriene C_4 production is increased in rejecting rat renal allografts. Kidney Int 29(I):332

Corey EJ, Niwa H, Falck JR, Mioskowski C, Arai Y, Marfat A, (1980) Recent studies on the chemical synthesis of eicosanoids. In: Samuelsson B, Ramwell P, Paoletti R (eds) Advances in prostaglandin and thromboxane research. Raven, New York, pp 19–25

Crowshaw K, Szlyk JZ (1970) Distribution of prostaglandins in rabbit kidney. Biochem J 116:421–424

Culpepper RM, Andreoli TE (1983) Interactions among prostaglandin E_2 anti-diuretic hormone and cyclic adenosine monophosphate in modulating Cl⁻ absorption in single mouse medullary thick ascending limbs of Henle. J Clin Invest 71:1588–1601

Daniel LW, King L, Waite M (1981) Source of arachidonic acid for prostaglandin synthesis in Madin-Darby canine kidney cells. J Biol Chem 256:12830–12835

Data JL, Chang LC, Nies AS (1976) Alteration of canine renal vascular response to hemorrhage by inhibitors of prostaglandin synthesis. Am J Physiol 230:940–945

Data JL, Gerber JG, Crump WJ, Frölich JC, Hollifield JW, Nies AS (1978) The prostaglandin system: a role in canine baroreceptor control of renin release. Circ Res 42:454–458

Davila D, Davila T, Oliw E, Anggard E (1978) Influence of dietary sodium on urinary prostaglandin excretion. Acta Physiol Scand 103:100–106

Davila D, Davila T, Oliw E, Anggard E (1980) Renal prostaglandins and sodium balance in the rabbit: lack of effect of aspirin-like drugs. Acta Pharmacol Toxicol 46:57–61

Davis JO, Freeman RH (1970) Mechanisms regulating renin release. Physiol Rev 56:1–56

DeForrest JM, Davis JO, Freeman RH, Seymour AA, Rowe BP, Williams GM, Davis TP (1980) Effects of indomethacin and meclofenamate on renin release and renal hemodynamic function during chronic sodium depletion in conscious dogs. Circ Res 47:99–107

Donker AJM, Arisz L, Brentjens JRH, van der Hem GK, Hollemans HJG (1976) The effect of indomethacin on kidney function and plasma renin activity in man. Nephron 17:288–296

Düsing R, Opitz WD, Kramer HJ (1977) The role of prostaglandins in the natriuresis of acutely salt-loaded rats. Nephron 18:212–219

Düsing R, Herrmann R, Kramer HT (1980) The renal prostaglandin system in central diabetes insipidus: effects of desamino-arginine vasopressin. In: Samuelsson P, Ramwell PW, Paoletti R (eds) Advances in Prostaglandin and Thromboxane Research vol 7. Raven, New York, pp 111–113

Düsing R, Gill JR Jr, Güllner H-G, Bartter FC (1982) The role of prostaglandins in diabetes insipidus produced by desoxycorticosterone in the dog. Endocrinology 110:644–649

Dunn MJ, Hood VL (1977) Prostaglandins and the kidney. Am J Physiol 233: F169–F184

Dunn MH, Greely HP, Valtin H, Kinter LB, Beeuwkes R III (1978a) Renal excretion of prostaglandins E_2 and F_{2a} in diabetes insipidus rats. Am J Physiol 235: 624–627

Dunn MJ, Liard JF, Dray F (1978b) Basal and stimulated rates of renal secretion and excretion of prostaglandins E_2, F_{2a}, and 13,14-dihydro-15-keto F_{2a} in the dog. Kidney Int 13:136–143

Edwards GA, Suki WN (1978) Effect of indomethacin on changes at acute ureteral pressure elevation in the dog. Renal Physiol 1:154–165

Edwards RM, Jackson BA, Dousa TP (1981) ADH-sensitive cAMP system in papillary collecting duct: effect of osmoality and PGE_2. Am J Physiol 240:311–318

Egan RW, Paxton J, Kuehl FA (1976) Mechanism for irreversible self-deactivation of prostaglandin synthetase. J Biol Chem 251:7329–7335

Eggena P, Schwartz L, Walter R (1970) Threshold and receptor reserve in the action of neurophypophyseal peptides: a study of synergists and antagonists of the hydroosmotic response of the toad urinary bladder. J Gen Physiol 56:250–271

Ellis CK, Smigel MD, Oates JA, Oelz O, Sweetman BJ (1979) Metabolism of prostaglandin D_2 in the monkey. J Biol Chem 254:4152–4163

Epstein M, Lifschitz MD (1980) Volume status as a determinant of the influence of renal PGE on renal function. Nephron 25:157–159

Epstein M, Lifschitz MD, Hoffmann DS, Stein HJ (1979) Relationship between renal prostaglandin E and renal sodium handling during water immersion on normal man. Circ Res 45:71–80

Epstein M, Lifschitz MD, Ramachandran M (1982) Characterization of renal prostaglandin E responsiveness in decompensated cirrhosis: implications for renal sodium handling. Clin Sci 63:555–563

Feigen LP (1983) Differential effects of leukotrienes C_3, D_4, and E_4 in the canine renal and mesenteric vascular beds. J Pharmacol Exp Ther 225:682–687

Feigen LP, Klainer E, Chapnick BM, Kadowitz PJ (1976) The effect of indomethacin on renal function in pentobarbital-anesthetized dogs. J Pharmacol Exp Ther 198:457–463

Fejes-Toth G, Szenasi G (1981) The effect of vasopressin on renal tubular ^{22}Na efflux in the rat. J Physiol Lond 318:1–7

Fejes-Toth G, Magyar, Al Walter J (1977) Renal response to vasopressin after inhibition of prostaglandin synthesis. Am J Physiol 232:F416–F423

Fejes-Toth G, Fekete A, Walter J (1978) Effect of antidiuretic hormone and indomethacin on intrarenal microsphere distribution. Pflugers Arch 376:67–72

Fejes-Toth G, Filep J, Mann V (1983a) Effect of vasopressin on prostaglandin excretion in conscious dogs. J Physiol Lond 344:389–397

Fejes-Toth G, Naray-Fejes-Toth A, Frölich JC (1983b) The acute effects of ADH on urinary prostaglandin excretion. J Pharmacol Exp Ther 227:215–219

Fejes-Toth G, Naray-Fejes-Toth A, Rigter B, Frölich JC (1983c) Urinary prostaglandin and kallikrein excretion are not flow dependent in the rat. Prostaglandins 25:99–103

Feldman D, Loose DS, Tan SY (1978) On-steroidal anti-inflammatory drugs cause sodium water retention in the rat. Am J Physiol 234:490–496

Ferreira SH, Vane JR (1967) Prostaglandins: their disappearance from and release into the circulation. Nature 216:868–873

Ferreri NR, Schwartzmann M, Ebrahan NG, Chender DN, McGiff JC (1984) Arachidonic acid metabolism in a cell suspension isolated from rabbit renal outer medulla. J Pharm Exp Ther 231:441–448

Feuerstein G, Feuerstein N (1980) The effect of indomethacin on isoprenaline-induced renin secretion in the cat. Eur J Pharmacol 61:85–88

Fichman MP, Littenburg G, Brooker G, Horton R (1972) Effect of prostaglandin A_1 on renal and adrenal function in man. Circ Res [Suppl II] 30, 32:11–19

Fichman M, Speckart P, Zia P, Lee A (1977) Antidiuretic response to ibuprofen in nephrogenic diabetes insipidus. Clin Res 25:165A

Filep J, Rigter B, Frölich JC (1984) Effect of synthetic leukotriene C_4 on renal electrolyte excretion in the conscious rat. In: Braquet P et al. (eds) Prostaglandins and membrane ion transport. Raven, New York, pp 281–285

Filep J, Rigter B, Frölich JC (1985) Vascular and renal effects of leukotrines C_4 in conscious rats. Am J Physiol 249:F739–F744

Fine LG, Trizna W (1977) Influence of prostaglandins on sodium transport of isolated medullary nephron segments. Am J Physiol 232:383–390

Fine LG, Kirschenbaum MA (1981) The direct effects of prostaglandins on sodium chloride transport in the mammalian nephron. Kidney Int 19:797–801

Finn WF, Arendshorst WJ (1976) Effect of prostaglandin synthetase inhibitors on renal blood flow in the rat. Am J Physiol 231:1541–1545

Fischer S, Weber PC (1984) Prostaglandin I_3 is formed in vivo in man after dietary eicosapentaenoic acid. Nature 307:165–168

Fischer S, Scherer B, Weber PC (1982) Prostacyclin metabolites, in urine of adults and neonates, studied by gas chromatography, mass spectrometry and radioimmunoassay. Biochem Biophys. Acta 710:493–501

Fitzgerald GAL, Friedman A, Miyamori I, O'Grady J, Lewis PJ (1979) A double blind placebo controlled crossover study of prostacyclin in man. Life Sci 25: 665–672

Flores AGA, Sharp GWG (1972) Endogenous prostaglandins and osmotic water flow in the toad bladder. Am J Physiol 223:1392–1397

Flower RJ (1974) Drugs which inhibit prostaglandin biosynthesis. Pharmacol Rev 26:33–67

Folkert VW, Schlondorf D (1979) Prostaglandin synthesis in isolated glomeruli. Prostaglandins 17:79–86

Forrest JN, Schneider CJ, Goodman DBP (1982) Role of prostaglandin E_2 in mediating the effects of pH on the hydroosmotic response to vasopressin in the toad urinary bladder. J Clin Invest 69:499–506

Francisco LL, Osborn JL, DiBona GF (1982) Prostaglandins in renin release during sodium deprivation. Am J Physiol 243:F537–F542

Freeman RH, Davis JO, Dietz JR, Villarreal D, Seymour AA, Echtemkamp SG (1982) Renal prostaglandins and the control of renin release. Hypertension 4 [Suppl II]: II-106 – II-112

Frölich JC (1977) Gas chromatography – mass spectrometry of prostaglandins. In: Ramwell P (ed) The prostaglandins, vol 3. Plenum, New York, pp 1–39

Frölich JC (ed) (1979) Methods in prostaglandin research. Raven, New York

Frölich JC, Sweetman BJ, Carr K, Oates JA (1975a) Prostaglandin synthesis in rabbit renal medulla. Life Sci 17:1105–1112

Frölich JC, Sweetman BJ, Carr K, Hollifield JW, Oates JA(1975) Assessment of the levels of PGA_2 in human plasma by gas chromatography – mass spectrometry. Prostaglandins 10:185

Frölich JC, Wilson TW, Sweetman BJ, Smigel M, Nies AS, Carr K, Watson JT, Oates JA (1975) Urinary prostaglandins: identification and origin. J Clin Invest 55: 763–770

Frölich JC, Hollifield JW, Oates JA (1976a) Effect of indomethacin on isoproterenol-induced renin release. Clin Res 24:9A

Frölich JC, Hollifield JW, Wilson JP, Sweetman BJ, Seyberth HW, Oates JA (1976b) Suppression of plasma renin activity in man by indomethacin: independence of sodium retention. Clin Res 24:397A

Frölich JC, Hollifield JW, Dormois JC, Frölich BL, Seyberth HJ, Michelakis AM, Oates JA (1976c) Suppression of plasma renin activity by indomethacin in man. Circ Res 39:447–452

Frölich JC, Whorton AR, Walker L, Smigel M, Oates JA, France R, Hollifield JW, Data JL, Gerber JG, Nies AS, Williams W, Robertson GL (1978) Renal prostaglandins regional differences in synthesis and role in renin release and ADH action. In: Karger S (ed) Proc of the VIIth international congress of nephrology. University of Montreal Press, Montreal, pp 107–114

Frölich JC, Gill JR, McGiff JC, Needleman P, Nies AS (1979a) Prostaglandins. Subgroup report of the hypertension task force. DHEW publiscation no (NIH) 79-1629. US Government Printing Office, Washington, pp 1–98

Frölich JC, Hollifield JW, Michelakis AM, Vesper BS, Shand DG, Wilson JP, Seyberth HJ, Frölich WH, Oates JA (1979b) Reduction of plasma renin activity by inhibition of the fatty acid cyclooxygenase in human subjects: independence of sodium retention. Circ Res 44:781–787

Frölich JC, Leftwich R, Ragheb M, Oates JA, Reimann I, Buchanan D (1979) Indomethacin increases plasma lithium. Br Med J 1:1115—1116

Frölich JC, Robertson D, Kitajima W, Rosenkranz B, Reimann I (1981) Prostaglandins in human hypertension: relationship to renin, sodium and antihypertensive drug action. In: Laragh JH, Bühler FR, Seldin DW (eds) Frontiers in hypertension research. Springer, Berlin Heidelberg New York, pp 114—118

Frölich JC (1981) Zur Rolle der Prostaglandine bei der Reninfreisetzung. Klin Wochenschr 59:1139—1147

Gagnon DJ, Cousineau D, Boucher PJ (1973) Release of vasopressin by angiotensin II and PGE_2 from the rat neurophypophysis in vitro. Life Sci 12:487—497

Gagnon JA, Felipe J (1979) Effect of prostaglandin synthesis inhibition on renal sodium excretion in the water-loaded conscious and anesthetized dog. Min Elect Metab 2:293—301

Galvez OB, Bay WH, Roberts BW, Ferris TF (1977) The hemodynamic effects of potassium deficiency in the dog. Circ Res 40 [Suppl I]:I-11 — I—16

Ganguli M, Tobian L, Azar S, O'Donell M (1977) Evidence that prostaglandin synthesis inhibitors increase the concentration of sodium and chloride in rat and renal medulla. Circ Res 40 [Suppl I]:135—139

Garcia-Perez A, Smith WL (1984) Apical-basolateral membrane assymetry in canine cortical collecting tubule cells: bradykinin, arginine vasopressin, prostaglandin E_2 interrelationships. J Clin Invest 74:63—74

Gaudio KM, Siegel NJ, Hayslett JP, Kashgarian M (1980) Renal perfusion and intratubular pressure during ureteral occlusion in the rat. Am J Physiol 238: F205—F209

Gerber JG, Nies AS (1980) Furosemide-induced vasodilation: importance of the state of hydration and filtration. Kidney Int 18:454—459

Gerber JG, Data JL, Nies AS (1978a) Enhances renal prostaglandin production in the dog: the effect of sodium arachidonate in the non filtering kidney. Circ Res 42:43—45

Gerber JG, Branch RA, Nies AS, Gerkens JF, Shand DG, Hollifield J, Oates JA (1978b) Prostaglandins and renin release II:Assessment of renin secretion following infusion of PGI_2, E_2 and D_2 into the renal artery of anesthetized dogs. Prostaglandins 15:81—88

Gerber JG, Nies AS, Friesinger GC, Gerkens JF, Branch RA, Oates JA (1978c) The effect of PGI_2 on canine renal function and hemodynamics. Prostaglandins 16:519—528

Gerber JG, Keller RT, Nies AS (1979) Prostaglandins and renin release. The effect of PGI_2, PGE_2 and 13,14-dihydro PGE_2 on the baroreceptor mechanism of release in the dog. Circ Res 44:796—799

Gerber JG, Olson RD, Nies AS (1981a) Interrelationships between prostaglandins and renin release. Kidney Int 19:816—821

Gerber JG, Nies AS, Data JL, Olsen RD (1981b) Control of canine renin release: macula densa requires prostaglandin synthesis. J Physiol 319:419—424

Gill JR, Frölich JC, Bowden RE, Taylor AA, Keiser HR, Seyberth H, Oates JA, Bartter FC (1976) Bartter's syndrome: a disorder characterized by high urinary prostaglandins and a dependence of hyperreninemia on prostaglandin synthesis. Am J Med 61:43—51

Gilmore N, Vane JR, Wyllie JH (1968) Prostaglandins released by the spleen. Nature 218:1135—1140

Goldblatt MW (1935) Properties of human seminal plasma. J Physiol (London) 84:208—218

Gordon RD, Kuchel O, Liddle GW, Island DP (1967) Role of the sympathetic nervous system in regulation of renin and aldosterone production in man. J Clin Invest 46:599—605

Gorman RR (1975) Prostaglandin endoperoxides: possible new regulators of cyclic nucleotide metabolism. J Cyclic Nuleotide Res 1:109

Granström E, Samuelsson B (1971) On the metabolism of PGF_{2a} in female subjects. J Biol Chem 246:7470–7485

Grantham JJ, Orloff J (1968) Effect of prostaglandin E_1 on the permeability response of the isolated collecting tubule to vasopressin, adenosine 3', 5'-monophosphate and theophylline. J Clin Invest 47:1154–1161

Grenier FC, Rollins TE, Smith WL (1981) Kinin-induced prostaglandin synthesis by renal papillary collecting tubule cells in culture. Am J Physiol 241:94–104

Grenier FC, Allen ML, Smith WL (1982) Interrelationships among prostaglandins, vasopressin and cAMP in renal papillary collecting tubule cells in culture. Prostaglandins 24:547–565

Gross DM, Mujovic VM, Jubiz W, Fisher JW (1976) Enhanced erythropoietin and prostaglandin E production in the dog following renal artery constriction. Proc Soc Exp Biol Med 151:498–501

Gross PA, Schrier RW, Anderson RJ (1981) Prostaglandins and water metabolism – a review with emphasis on in vivo studies. Kidney Int 19:839–850

Grosso A, Cox JA, Malnoe A, deSousa RC (1982) Evidence for a role of calmodulin in the hydroosmotic action of vasopressin in toad bladder. J Physiol 78:270–278

Gryglewski RJ (1979) The lung as an endocrine organ. In: Vane JR, Bergström S (eds) Prostacyclin. Raven, New York, pp 275–288

Güllner HG, Gill JR, Bartter FC, Düsing R (1980a) The role of the prostaglandin system in the regulation of renal function in normal women. Am J Med 69: 718–724

Güllner HG, Nicolaou KC, Bartter FC, Kelly G (1980b) Effect of prostacyclin (PGI_2) on renal function and renin secretion in hypophysectomized dogs. Nephron 25:283–287

Hall DA, Varney DM (1980) Effect of vasopressin on electrical potential difference and chloride transport in mouse medullary thick ascending limb of Henle's loop. J Clin Invest 66:792–802

Halliday HL, Hirata T, Brady JP (1979) Indomethacin therapy for large patent ductus arteriosus in the very low birth weight infant: results and complications. Rediatrics 64:154–159

Hamberg M, Fredholm BB (1976) Isomerization of prostaglandin H2 into prostaglandin D_2 in the presence of serum albumin. Biochim Biophys. Acta 431: 189–193

Hamberg M, Israelsson U (1970) Metabolism of prostaglandin E_2 in guinea pig liver. I. Identification of seven metabolites. J Biol Chem 245:5107–5144

Hamberg M, Samuelsson B (1966) Prostaglandins in human seminal plasma. J Biol Chem 241:257–263

Hamberg M, Samuelsson B (1971) On the metabolism of PGE_1 and PGE_2 in man. J Biol Chem 246:6713–6721

Hamberg M, Samuelsson B (1973) Detection and isolation of an endoperoxide intermediate in prostaglandin biosynthesis. Proc Natl Acad Sci USA 70:899–903

Hamberg M, Samuelsson B (1974) Prostaglandin endoperoxides VII. Novel transformations of arachidonic acid in guinea pig lung. Biochim Biophys Res Commun 61:942–949

Hamberg M, Sevensson J, Samuelsson B (1974) Prostaglandin endoperoxides: Novel transformations of arachidonic acid in human platelets. Proc Natl Acad Sci USA 71:3400–3404

Hammarström S (1980) Leukotriene C_5: a slow reacting substance derived from eicosapentaenoic acid. J Biol Chem 255:7093–7094

Hammarström S (1983) Leukotrienes. Annu Rev Biochem 52:355–377

Hardy MA (1978) Intracellular calcium as a modulator of transepithelial permeability of water in frog urinary bladder. J Cell Biol 76:787–789

Harris RN, Yarger WE (1974) Renal function after release of unilateral ureteral obstruction in rats. Am J Physiol 227:806–815

Hassid A, Dunn MJ (1982) Biosynthesis and metabolism of prostaglandins in human kidney in vitro. In: Dunn MJ, Patrono C, Cinotti GA (eds) Prostaglandins and the kidney. Plenum, New York, pp 3–15

Haye B, Champion S, Jacquemin C (1973) Control by TSH of a phospholipase A_2 activity, a limiting factor in the biosynthesis of prostaglandins in the thyroid. FEBS Lett 30:253–260

Haylor J (1980) Prostaglandin synthesis and renal function in man. J Physiol 298: 383–396

Haylor J, Lote CJ (1980) Renal function in conscious rats after indomethacin. Evidence for a tubular action of endogenous prostaglandins. J Physiol 298: 371–381

Haylor J, Lote CJ (1983) The influence of prostaglandin E_2 and indomethacin on the renal corticomedullary solute gradient in the rat. J Pharm Pharmacol 35: 298–305

Haylor J, Towers J (1982) Renal vasodilator activity of prostaglandin E_2 in the rat anaesthetized with pentobarbitone. Br J Pharmacol 76:131–137

Haylor J, Lote CJ, Thewles A (1984) Urinary pH as a determinant of prostaglandin E_2 excretion by the conscious rat. Clin Sci 66:675–681

Hebert SC, Andreoli TE (1984) Control of NaCl transport in the thick ascending limb. Am J Physiol 246:F745–F756

Hember M, Lands WEM, Smith WL (1975) Purification of the cyclooxygenase that forms prostaglandins. J Biol Chem 251:5575–5579

Henrich WL (1981) Role of prostaglandins in renin secretion. Kidney Int 19:822–830

Henrich WL, Campbell WB (1983) The systemic beta-adrenergic pathway to renin secretion: relationship with the renal prostaglandin system. Endocrinology 113:2247–2254

Henrich WL, Anderson RJ, Berns AS, McDonald KM, Paulsen PJ, Berl T, Schrier RW, (1978a) Role of renal nerves and prostaglandins in control of renal hemo-dynamics and plasma renin activity during hypotensive hemorrhage in dogs. J Clin Invest 61:744–750

Henrich WL, Berl T, McDonald KM, Anderson RJ, Schrier RA (1978b) Angiotensin II, renal nerves and prostaglandins in renal hemodynamics during hemorrhage. Am J Physiol 235:F46–F51

Henrich WL, Schrier RW, Berl T (1979) Mechanisms of renin secretion during hemor-rhage in the dog. J Clin Invest 64:1–7

Herbaczynska-Cedro K, Vane JR (1973) Contribution of intrarenal generation of prostaglandin to autoregulation of renal blood flow in the dog. Circ Res 33: 428–436

Herman CA, Zenser V, Davis BB (1979) Comparison on the effects of prostaglandin I_2 and prostaglandin E_2 stimulation of the rat kidney adenlyate cyclase-cyclic AMP systems. Biochim Biophys Acta 582:496

Herman CA, Zenser T, Davis B (1980) Effects of prostaglandin H_2, prostaglandin E_2 and arachiconid acid on parathyroid hormone and antidiuretic hormone activation of rat kidney adenylate cyclase. Metabolism 29:1–8

Higashihara E, Stokes JB, Kokko JP, Campbell WB, DuBose TD Jr (1979) Cortical and papillary micropuncture examination of chloride transport in segments of the rat kidney during inhibition of prostaglandin production: a possible role of prostaglandins in the chloruresis of acute volume expansion. J Clin Invest 64:1277–1287

Hirata F, Axelrod J (1980) Phospholipid methylation and biological signal tans-mission. Science 209:1082–1090

Holt WF, Lechene C (1981a) ADH-PGE_2 interactions in cortical collecting tubule. I. Depression of sodium transport. Am J Physiol 241:F452–F460

Holt WF, Lechene C (1981b) ADH-PGE_2 interactions in cortical collecting tubule II: inhibition of Ca and P reabsorption. Am J Physiol 241:F461–F467

Hong SL, Deykin D (1981) The activation of phosphatidylinositol hydrolyzing phospholipase A_2 during prostaglandin synthesis in transformed mouse BALB/3T3 cell. J Biol Chem 256:5215-5219

Hood VL, Dunn MJ (1978) Urinary excretion of PGE_2 and $PGF_{2\alpha}$ in potassium deficient rats. Prostaglandins 15:273-283

Hoult JRS, Moore PK (1977) Pathways of prostaglandin $F_{2\alpha}$-metabolism in mammalian kidneys. Br J Pharmacol 61:615-626

Humes DM, Simmon CF, Brenner BM (1980) Effect of verapamil in the hydroosmotic response to antidiuretic hormone in the toad urinary bladder. Am J Physiol 239:F250-F251

Ichikawa I, Brenner BM (1979) Local intrarenal vasocontriction-vasodilator interactions in mild partial ureteral obstruction. Am J Physiol 236:F131-F140

Lino Y, Imai M (1978) Effects of prostaglandins on Na transport in isolated collecting tubules. Pflügers Arch 373:125-132

Imbs JL, Singer L, Danion JM, Schmidt M, Sebban M (1979) Urinary elimination of lithium: drug interactions. Kidney Int 16:96

Inokiechi K, Malik KU (1984) Attenuation by prostaglandins of adrenergically induced renal vasoconstriction in anesthetized rats. Am J Physiol 246:R228-R235

Itskovitz HD, Stemper J, Pacholczyk D, McGiff JC (1973) Renal prostaglandins: determinants of intrarenal distribution of blood flow in the dog. Clin Sci Mol Med 45:321s-324s

Itskovitz HD, Terragno NA, McGiff JC (1974) Effect of a renal prostaglandin on distribution of blood flow in the isolated canine kidney. Circ Res 34:770-776

Jackson BA, Edwards RM, Dousa TP (1980) Vasopressin − prostaglandin interactions in isolated tubules from rat outer medulla. J Lab Clin Med 96:119

Jackson EK, Gerkens JF, Brash AR, Branch RA (1982a) Acute renal artery constriction increases renal prostaglandin I_2· biosynthesis and renin release in the conscious dog. J Pharmacol Exp Ther 222:410-413

Jackson EK, Heidemann HT, Branch RA, Gerkens JF (1982b) Low dose intrarenal infusions of PGE_2, PGI_2 and 6-keto-PGE_1 vasodilate the in vivo rat kidney. Cirs Res 51:67-72

Jacobson HR, Corona S, Capdeville J, Chacos N, Manna S, Womack A, Falck JR (1984) 5, 6 epoxycosatrienoic acid inhibits sodium absorption and potassium secretion in rabbit cortical collecting tubule (abstract). Kidney Int 25:330

Jakschik BA, Marshall GR, Lourik JL, Needleman P (1974) Profile of circulating vasoactive substances in hemorrhagic shock and their pharmacologic manipulation. J Clin Invest 54:842-852

Janszen FH, Nugteren DH (1971) Histochemical localization of prostaglandin synthetase. Histochemistry 27:159-164

Jim K, Hassid A, Sun F, Dunn MJ (1982) Lipoxygenase activity in rat kidney glomeruli, glomerular epithelial cells and cortical tubules. J Biol Chem 257:10294-10299

Johnson KL, Morton DR, Kinner JH, Gorman RR, McGuire JC, Sun FF, Whittacker N, Bunting S, Salomon J, Moncada S, Vane JR (1976) The chemical structure of prostaglandin X (Prostacycline). Prostaglandins 12:915-928

Johnston HH, Herzog JP, Lauler DP (1967) Effect of prostaglandin E_1 on renal hemodynamics, sodium and water excretion. Am J Physiol 213:939-946

Kadokawa T, Hosoki K, Takeyama K, Minato H, Shimizu M (1979) Effects of nonsteroidal anti-inflammatory (NSAID) on renal excretion of sodium and water, and on body fluid volume in rats. J Pharmacol Exp Ther 209:219-224

Kalisker A, Dyer DC (1972) In vitro release of prostaglandins from the renal medulla. Eur J Pharmacol 19:305-309

Kaloyanides GJ, Ahrens RE, Shepherd JA, DiBona GF (1976) Inhibition of prostaglandin E_2 secretion: failure to abolish autoregulation in the isolated dog kidney. Circ Res 38:67-73

Kaojarern S, Chennarosin P, Anderson S, Brater DC (1983) Nephron site of effect of non-steroidal anti-inflammatory drugs on solute excretion. Am J Phsiol 244:F134–F139

Kauker ML (1977) Prostaglandin E_2 effect from the luminal side on renal tubular. [22] Na efflux: tracer microinjection studies. Proc Soc Exp Biol Med 154:274–277

Kaye Z, Zipser RD, Mayeda S, Zia PK, Horton R (1979) Renal prostaglandins and sodium balance in normal man. Prostaglandins Med 2:123–131

Kaye Z, Zipser R, Hahn J, Zia P, Horton R (1980) Water diruesis is a major regulator of prostaglandin E excretion in man. Adv Prostaglandin Thromboxane Res 7:1017–1019

Kimberley RP, Plotz PH (1977) Aspirin-induced depression of renal function. N Engl J Med 296:418–424

Kimberley RP, Gill JR, Bowden RE, Keiser HR, Plotz PH (1978) Elevated urinary prostaglandins and the effects of aspirin on renal function in lupus erythematosus. Ann Intern Med 89:336–341

Kirschenbaum MA, Serros ER (1980a) Effects of alterations in urine flow rate on prostaglandin E excretion in conscious dogs. Am J Physiol 238:107–111

Kirschenbaum MA, Serros ER (1980b) Are prostaglandins natriuretic? Miner Electro-lyte Metab 3:113–121

Kirschenbaum MA, Stein JH (1976) Effect of inhibition of prostaglandin synthesis on urinary sodium excretion in conscious dog. J Clin Invest 57:517–521

Kirschenbaum MA, White N, Stein JH, Ferris TG (1974) Redistribution of renal cortical blood flow during inhibition of prostaglandin synthesis. Am J Physiol 227:801–805

Kirschenbaum MA, Lowe AG, Trizna W, Fine LG (1982) Regulation of vasopressin action by prostaglandins. J Clin Invest 70:1193–1204

Knapp HR, Oelz O, Whorton AR, Oates JA (1978) Effects of feedings ethyl-dihomo-y-linolenate on rabbit renomedullary lipic composition and prostaglandin production in vitro. Lipids 13:804–808

Kokko JP (1981) The effect of prostaglandins on renal epithelial electrolyte transport. Kidney Int 19:791–796   •

Kopp U, Aurell M, Sjolander M, Ablad B (1981) The role of prostaglandins in the alpha- and beta-adrenoceptor mediated renin release response to graded renal nerve stimulation. Pflugers Arch 391:1–8

Krakoff LR, Deguia D, Vlachakis N, Stricker J, Goldstein M (1973) Effect of sodium balance on arterial blood pressure and renal responses to PGA_1 in man. Circ Res 33:539

Kramer HJ, Backer A, Hinzen S, Dusing R (1978) Effects of inhibition of prosta-glandin-synthesis on renal electrolyte excretion and concentrating ability in healthy man. Prostaglandins Med 1:341–349

Kramer HJ, Glänzer K, Düsing R (1981) The role of prostaglandins in the regulation of renal water excretion. Kidney Int 19:851–859

Kreisberg JI, Karnovsky MJ, Levine L (1981) Prostaglandin production by homo-genous cultures of rat glomerular epithelial and mesongial cells. Kidney Int 22:355–359

Kunze H, Vogt W (1971) Significance of phospholipase A for prostaglandin forma-tion. Ann NY Acad Sci 180:123–125

Kurtz A, Jelkmann W, Sinowatz F, Bauer C (1983) Renal mesangial cells in culture in a model for study of erythropoietin production. Proc Natl Acad Sci USA 80:4008–4011

Kurtz A, Jelkmann W, Pfeilschifter J, Bauer C (1985) Role of prostaglandins in hypoxia-stimulated erythropoietin production. Am J Physiol 249:C3–C8

Kurzrok R, Lieb CC (1930) Biochemical studies of human semen. II. The action of semen of the human uterus. Proc Soc Exptl Biol Med 28:268–272

Lameire N, Ringoir S, Leusen I (1980) The effect of prostaglandin synthesis in-
    hibitors on sodium excretion in the awake rat during acute water and isotonic
    Ringer loading. Arch Int Pharmacodyn. Ther 244:141—156
Lands WEM, Samuelsson B (1968) Phospholipid precursors of prostaglandins. Biochim
    Biophys Acta 164:426—429
Larsson C, Anggard E (1973) Regional differences in the formation and metabolism
    of prostaglandins in the rabbit kidney. Eur J Pharmacol 21:30—36
Larsson C, Anggard E (1974) Increased juxtamedullary blood flow on stimulation
    of intrarenal prostaglandin biosynthesis. Eur J Pharmacol 25:326—334
Larsson C, Weber P, Anggard E (1974) Arachidonic acid increases and indomethacin
    degreases plasma renin activity in the rabbit. Eur J Pharmacol 28:391
Lee JB, Covino BJ, Takman BH, Smith ER (1965) Renomedullary vasodepressor
    substance medullin: isolation, chemical characterization, and physiological
    properties. Circ Res 17:57—70
Lee JB, Crowshaw K, Takman GH, Attrep KA, Gougoutas JZ (1967) The identifica-
    tion of prostaglandin E_2, F_{2a} and A_2 from rabbit kidney medulla. Biochem
    J 105:1251—1260
Lee SC, Levine L (1974) Prostaglandin metabolism. I. Cytoplasmic reduced incotin-
    amide adenine nucleotide phosphate-dependent and microsomal reduced nicotin-
    amide adenine dinucleotide dependent prostaglandin 9-keto-reductase activities
    in monkey and pigone tissues. J Biol Chem 249:1369—1375
Lefer AM (1986) Leukotrienes as mediators of ischemia and shock. Biochem Phar-
    macol 35(2):123—127
Lemley KV, Schmitt SF, Hollinger C, Dunn MJ, Robertson CR, Jamison RL (1984)
    Prostaglandin synthesis inhibitors and vasa recta erythrocyte velocities in the rat.
    Am J Physiol 247:F562—F567
Levine SD, Kachadorian WA, Levin DN, Schlondorff D (1981) Effects of triflu-
    operacine on function and structure of toad urinary bladder: the role of cal-
    modulin in vasopressin stimulation of water permeability. J Clin Invest 61:
    662—672
Levy M, Wexler MJ, Fechnet C (1983) Renal perfusion in dogs with experimental
    hepatic cirrhosis: role of prostaglandins. Am J Physiol 245:F521—529
Lewis RA, Austen KF (1984) The biologically active leukotrienes: biosynthesis,
    metabolism, receptors, functions, and pharmacology. J Clin Invest 73:889—897
Leyssac PP, Christensen P (1981) On the relationship between urinary PGE_2 and
    PGF_{2a} excretion rates and urine flow, osmolar excretion rate and urinary
    osmoality in anesthetized rats. Acta Physiol Scand 113:427—435
Leyssac PP, Christensen P, Hill R, Skinner SL (1975) Indomethacin blockade of renal
    PGE-synthesis: effect on total renal and tubular function and plasma renin
    concentration in hydropenic rats and on their response to isotonic saline. Acta
    Physiol Scand 94:484—496
Lianos EA (1986) Glomerular leukotriene (LT) biosynthesis and degradation in the
    rat: Effects of immune injury. Kidney Int 29(1):339
Licht A, Alpert BE, Bourgoignie JJ, Schlondorff D (1981) The effect of prostaglandin
    synthetase inhibitors on renal function: studies in normal rats during sodium or
    water diuresis and in rats with chronic renal insufficiency. Prostaglandins 22:
    1—10
Lieberthal W, Levine L (1984) Stimulation of prostaglandin production in rat
    glomerular epithelial cells by antidiuretic hormone. Kidney Int 25:766—770
Lifschitz M, Stein JH (1977) Antidiuretic hormone stimulates renal prostaglandin E
    (PGE) synthesis in the rabbit. (abstract) Clin Res 25:440A
Lifschitz MD, Patak RV, Fadem SZ, Stein JH (1978) Urinary prostaglandin E ex-
    cretion. Effect of chronic alterations in sodium intake and inhibition of prosta-
    glandin synthesis in the rabbit. Prostaglandins 16:607--619

Lin CS, Iwao H, Puttkammer S, Michelakis AM (1981) Prostaglandins and renin release in vitro. Am J Physiol 240:E609–E614

Linas SL (1984) Role of prostaglandins in renin secretion in the isolated kidney. Am J Physiol 246:F811–F818

Linas SL, Dickmann D (1982) Mechanism of the decreased renal blood flow in the potassium-depleted conscious rat. Kidney Int 21:757–764

Lipson LC, Sharp GW (1971) Effect of prostaglandin E_1 on sodium transport and osmotic water flow in toad bladder. J Physiol 220:1046–1052

Logan A, Jose P, Eisner G, Lilienfield L, Slotkoff L (1971) Intracortical distribution of renal blood flow in hemorrhagic shock in dogs. Circ Res 29:257–266

Lohmeier TE, Cowley AW Jr Trippodo NC, Hall JE, Guyton AC (1977) Effects of endogenous angiotensin II on renal sodium excretion and renal hemodynamics. Am J Physiol 233:388–395

Lonigro J, Itskovitz D, Crowshaw K, McGiff C (1973) Dependency of renal blood flow on prostaglandin synthesis in the dog. Circ Res 32:712–717

Ludens JH, Hook JB, Brody MJ, Williamson HE (1980) Enhancement of renal blood flow by furosemide. J Pharmacol Exp Ther 163:456–460

Lum GM, Aisenbrey A, Dunn MJ, Berl T, Schrier RW, McDonald KM (1977) In vivo effect of indomethacin to potentiate the renal medullary cyclic AMP response to vasopressin. J Clin Invest 59:8–13

Malik KU, McGiff JC (1975) Modulation by prostaglandins of adrenergic transmission in the isolated perfused rabbit and rat kidney. Circ Res 36:599–609

Mann S, Falck JR, Chacos N, Capdevilla J (1983) Synthesis of arachidonic acid metabolites produced by purified kidney cortex microsomal cytochrome P-450. Tetra Let 24:33–36

Martinez F, Reyes JL (1984) Prostaglandin receptors and hormonal actions on water fluxes in cultured canine renal cells (MDCK line). J Physiol 347:533–543

Martinez-Maldonado M, Saparas N, Eknoyan G, Suki WN (1972) Renal actions of prostaglandins: comparison with acetylcholine and volume expansion. Am J Physiol 222:1147–1152

Marumo F, Edelman IE (1971) Effects of Ca^{++} and prostaglandin E_1 on vasopressin activation of renal adenyl cyclase. J Clin Invest 50:1613–1620

McGiff JC, Wong PYK (1979) Compartmentalization of prostaglandins and prostacyclin with in the kidney. Implications for renal function. Fed Proc 38:89–93

McGiff JC, Terragno NA, Strand JC, Lee JC, Lonigro AJ, Ng KKF (1969) Selective passage of prostaglandins across the lung. Nature 223:742–745

McGiff JC, Crowshaw K, Terragno NA, Lonigro AJ (1970) Release of prostaglandin-like substance into renal venous blood in response to angiotensin II. Circ Res 27 [Suppl I]:121–130

McGiff JC, Spokas EG, Wong PYK (1982) Stimulation of renin release by 6-oxo-prostaglandin E_1 and prostacyclin. Br J Pharmacol 75:137–144

Mimran A, Casellas D, DuPont M, Baron P (1975) Effect of competitive angiotensin antagonist on the renal hemodynamic changes induced by inhibition of prostaglandin synthesis in rats. Clin Sci Mol Med 48:299s–302s

Miyamoto T, Ogino N, Yamamoto S, Hayaishi O (1976) Purification of prostaglandin endoperoxide synthetase from bovine vesicular gland microsomes. J Biol Chem 251:2629–2636

Moncado S, Gryglewski R, Bunting S, Vane JR (1976) An enzyme isolated from arteries transforms prostaglandin endoperoxides to an unstable substance that inhibits platelet aggregation. Nature 263:663–665

Montgomery SB, Jose PA, Slotkoff LM, Lilienfield LS, Eisner GM (1980) The regulation of intrarenal blood flow in the dog during ischemia. Circ Shock 7:71–82

Moody TE, Vaughn ED, Gillenwater JY (1975) Relationship between renal blood flow and ureteral pressure during eighteen hours of total unilateral occlusion. Implications for changing sites of renal resistance. Invest Urol 31:246–251

Morel F, Imbert-Teboul M, Chabardes D (1980) Cyclic nucleotides and tubule function. Adv Cyclic. Nucleotide Res 12:301

Morrison AR, Pascoe N (1981) Metabolism of arachidonic acid through NADPH-dependent oxygenase of renal cortex. Proc Natl Acad Sci USA 78:7375–7378

Morrison AR, Nishikawa K, Needleman P (1977) Unmasking of thromboxane A_2 synthesis by ureteral obstruction in the rabbit kidney. Nature 267:295–260

Morrison AR, Nishikawa K, Needleman P (1978) Thromboxane A_2 biosynthesis in the ureter obstructed isolated perfused kidney of the rabbit. J Pharmacol Exp Ther 205:1–8

Mountokalakis T, Karambasis T, Mayopoulou-Symvoulidou D, Merikas G (1978) Effect of inhibition of prostaglandin synthesis on the natriuresis induced by saline infusion in man. Clin Sci Mol Med 54:47–50

Muther RS, Potter DM, Bennett WM (1981) Aspirin-induced depression of glomerular filtration rate in normal humans: role of sodium balance. Ann Intern Med 94:317–321

Nadler J, Zipser RD, Coleman R, Horton R (1983) Stimulation of renal prostaglandin by pressor hormones in man: Comparison of prostaglandin E_2 and prostacyclin. J Clin Endocrinol Metab 56:1260–1265

Needleman P, Raz A, Ferrendelli JA, Minkes M (1977) Application of imidazole as a selective inhibitor of thromboxane synthesis in human platelets. Proc Natl Acad Sci USA 74:1716–1720

Nugteren DH, Hazelhof E, (1973) Isolation and properties of intermediates in prostaglandin biosynthesis. Biochim Biophys Acta 326:448

Nugteren DH, van Dorp DA (1965) The participation of molecular oxygen in biosynthesis of prostaglandins. Biochim Biophys Acta 98:654–656

Nusynowitz ML, Forsham PH (1966) The antidiuretic action of acetaminophen. Am J Med Sci 252:429–435

Olgetree ML, Heran CL (1985) Inhibition of leukotriene C_4-induced changes in renal blood flow and hematocrit by a selective thromboxane antagonist, SQ 29548. (abstract) Pharmacologist 27:176

Okahara T, Imanishi M, Yamamoto K (1983) Zonal heterogeneity of prostaglandin and thromboxane release in the dog kidney. Prostaglandins 25:373–383

Okegawa T, Jonas PE, DeSchryvev K, Kawasaki A, Needleman P (1983) Metabolic and cellular alterations underlying the exaggerated renal prostaglandin and thromboxane synthesis in ureter obstruction in rabbits. J Clin Invest 71:81–90

Oliw EH, Oates JA (1981) Rabbit renal cortical microsomes metabolize arachidonic acid to trihydroxyeicosatrienoic acids. Prostaglandins 22:863–871

Oliw E, Kover G, Larsson C, Anggard E (1978) Indomethacin and diclofenac sodium increase sodium and water excretion after extracellular volume expansion in the rabbit. Eur J Pharmacol 49:381–388

Oliver JA, Pinto J, Sciacca RR, Cannon PJ (1980) Increased renal secretion of norepinephrine and prostaglandin E_2 during sodium depletion in the dog. J Clin Invest 66:748–756

Oliver JA, Sciacca RR, Pinto J, Cannon JA (1981) Participation of prostaglandins in the control of renal blood flow during actute reduction of cardiac output in the dog. J Clin Invest 67:229–237

Olsen UB, Ahnfelt-Ronne I (1976) Renal cortical blood redistribution after bumetanide related to heterogenicity or cortical prostaglandin metabolism in dogs. Acta Physiol Scand 97:251–257

Olsen UB, Magnussen MP, Eilertsen E (1976) Prostaglandins, a link between renal hydro- and hemodynamic in dogs. Acta Physiol Scand 97:369–376

Olson RD, Nies AS, Gerber JG (1981) Alpha adrenergic-mediated renin release is prostaglandin-dependent. J Pharmacol Exp Ther 219:321–325

Omachi RS, Robbie DE, Handler JS, Orloff J (1974) Effects of ADH and other agents on cyclic AMP accumulation in toad bladder epithelium. Am J Physiol 226:1152–1157

Orloff J, Handler JS, Bergstrom S (1965) Effect of prostaglandin (PGE_1) on the permeability response of toad bladder to vasopressin, theophylline and adenosine 3', 5'-monophosphate. Nature 205:397–398

Osgood RW, Reineck HJ, Stein JH (1978) Further studies on segmental sodium transport in the rat kidney during expansion of the extracellular fluid volume. J Clin Invest 62:311–320

Owen TL, Ehrhart I'Widner WJ, Scott JB, Haddy FJ (1975) Effects onf indomethacin on local blood flow regulation in canine heart and kidney. Proc Soc Exp Biol Med 149:871–876

Ozer A, Sharp G (1982) Effects of prostaglandins and their inhibitors on osmotic water flow in the toad bladder. Am J Physiol 22:674–680

Pace-Asciak C (1975) Prostaglandin 9-hydroxy-dehydrogenase activity in the adult rat kidney: identification, assay, pathway, and some enzyme properties. J Biol Chem 250:2789–2794

Pace-Asciak C (1976a) Isolation, structure and biosynthesis of 6-keto-prostaglandin F_{1a} in the rat stomach. J Am Chem Soc 98:2348–2349

Pace-Asciak C, Wolfe LS (1971) A novel prostaglandin derivative formed from arachidonic acid by rat stomach homogenates. Biochemistry 10:3657–3664

Pace-Asciak C, Nashat M, Menon NK (1976) Transformation of prostaglandin G_2 into 6(9) oxy-11, 15-dihydroxy-prosta-7, 13-dienoic acid by the rat stomach fundus. Biochim Biophys Acta 424:323–325

Palmer MA, Piper PJ, Vane JR (1973) Release of rabbit aorta contracting substance (RCS) and prostaglandins induced by chemical or mechanical stimulation of guinea-pig lungs. Br J Pharmacol 49:226–242

Parisi M, Piccini ZF (1972) Aspirin potentiates the hydroosmotic effect of anti-diuretic hormone in toad urinary bladder. Biochim Biophys Acata 279:209–219

Patak RV, Mookerjee BK, Bentzel CJ, Hysert PW, Babej M, Lee JB (1975) Antagonism of the effects of furosemide by indomethacin in normal and hypertensive man. Prostaglandins 10:649–659

Patrono C, Mennmalm A, Ciabattoni G, Nowak J, Pugliese F, Cinotti GA (1979) Evidence for an extra-renal origin of urinary prostaglandin E_2 in healthy men. Prostaglandins 18:623–629

Patrono C, Pugliese F, Ciabottoni G, Patrignani P, Maseri A, Chierchia S, Peskar BA, Cinotti GA, Simonetti BM, Pierucci A (1982) Evidence for a direct stimulatory effect of prostacyclin on renin release in man. J Clin Invest 69:231–239

Peterson LN, Gerber JF, Nies AS (1984) Effect of pH on the permeability of the distal nephron to prostaglandins E_2 and F_2. Am J Physiol 246:F221-F226

Petrulis A, Aikawa M, Dunn MJ (1981) Prostaglandin and thromboxane synthesis by rat glomerular epithelial cells. Kidney Int 20:469–474

Pong SS, Levine L (1976) Biosynthesis of prostaglandins in rabbit renal cortex. Res Commun Chem Pathol Pharmacol 13:115–123

Quereshi Z, Cagen LM (1982) Prostaglandin F_{2a} produced by rabbit renal slices is not a metabolite of prostaglandin E_2. Biochem Biophys Res Commun 104:1255–1263

Ragheb M, Ban TA, Buchanan D, Frölich JC (1980) Interaction of indomethacin and ibuprofen with lithium in manic patients under a steady state lithium level. J Clin Psychiatry 41:11

Ramwell PW, Shaw JE (1970) Biological significance of the prostaglandins. Recent Prog Hormon Res 26:139–173

Ray C, Morgan T (1981) The effect of prostaglandin E_2 and ADH on diffusional water permeability in the collecting duct of an isolated rat papilla. Pflügers Arch 392:51–56

Rector JB, Stein JH, Bay WH, Osgood RW, Ferris TF (1972) Effect of hemorrhage and vasopressor agents on distribution of renal blood flow. Am J Physiol 222:1125–1131

Reimann IW, Frölich JC (1981) Effects of diclofenac on lithium kinetics. Clin Pharmacol Ther 30:348–352

Reimann IW, Diener U, Frölich JC (1983a) Indomethacin but not aspirin increases plasma lithium levels. Arch Gen Psychiatry 40:283–287

Reimann IW, Fischer C, Rosenkranz B, Frölich JC (1983b) Investigations on renal prostaglandins by gas chromatography – mass spectrometry. In: Dunn MJ, Patrono C, Cinotti GA (eds) Prostaglandins and the kidney. Plenum, New York, pp 99–107

Rocha AS, Kokko JP (1974) Permeability of medullary nephron segments to urea and water: effect of vasopressin. Kidney Int 6:379–387

Rodriguez JA, Delea CS, Bartter FC, Siragy H (1981) The effect of vasopressin in water-loaded hypokalemic patients is prostaglandin independent. Prostaglandin Med 7:465–472

Roman RJ, Kauker ML (1978) Renal effect of prostaglandin synthetase inhibition in rats: Micropuncture studies. Am J Physiol 235:111–118

Roman RJ, Lechene C (1981) Prostaglandin E_2 and F_{2a} reduce urea reabsorption from the rat collecting duct. Am J Physiol 241:53–60

Romero JC, Dunlap CL, Strong CG (1976) The effect of indomethacin and other anti-inflammatory drugs on the renin-angiotensin system. J Clin Invest 58:282

Rosenblatt SG, Patak RV, Lifschitz MD (1978) Organic acid secretory pathway and urinary excretion of prostaglandin E in the dog. Am J Physiol 235:F473–F479

Rosenkranz B, Fischer C, Weimer KE, Frölich JC (1980) Metabolism of prostacyclin and 6-keto-prostaglandin F_{1a} in man. J Biol Chem 255:10194–10198

Rosenkranz B, Kitajima W, Frölich JC (1981a) Relevance of urinary 6-keto-prostaglandin F_{1a} determination. Kidney Int 19:759

Rosenkranz B, Fischer C, Frölich JC (1981b) Prostacyclin metabolites in human plasma. Clin Pharmacol Ther 29:420–424

Rosenkranz B, Wilson TW, Seyberth H, Frölich JC (1981c) Prostaglandins and renal blood flow. Proceedings, VIIIth international congress of nephrology, Athens, 1981. In: Advances in basic and clinical nephrology. Karger, Basel, pp 1045–1052

Rosenthal A, Pace-Asciak CR (1983) Potent vasoconstriction of the isolated perfused rat kidney by leukotrienes C_4 and D_4. Can J Physiol Pharmacol 61:325–328

Rutecki GW, Cox JW, Robertson GW, Francisco LL, Ferris TF (1982) Urinary concentrating ability and antidiuretic hormone responsiveness in the potassium-depleted dog. J Lab Clin Med 1:53–60

Ryhage R, Samuelsson B (1965) The origin of oxygen incorporated during the biosynthesis of prostaglandin E_1. Biochem Res Commun 19:279–282

Sakr HM, Durham EW (1982) Mechanism of arachidonic acid-induced vasoconstriction in the intact rat kidney: possible involvement of thromboxone A_2. J Pharmacol Exp Ther 221:614–622

Samuelsson B (1963) Prostaglandins and related factors, 17: the structure of prostaglandin E_3. J Am Chem Soc 85:1878–1879

Samuelsson B (1964a) Identification of a smooth muscle stimulating factor in bovine brain: prostaglandins and related factors, 25. Biochim Biophys Acta 84:218–219

Samuelsson B (1964b) Identification of prostaglandin F_{3a} in bovine lung prostaglandins and related factors, 26. Biochim Biophys Acta 84:707–713

Samuelsson B (1965) On the incorporation of oxygen in the conversion of 8, 11, 14-eicosatrienoic acid to prostaglandin E_1. J Am Chem Soc 87:3011–3013

Samuelsson B (1972) Biosynthesis of prostaglandins. Fed Proc 31:1442–1450

Samuelsson B (1973) Quantitative aspects of prostaglandin synthesis in man. First international congress on prostaglandins. In: Bergström S, Bernhard S (eds) Advances in the biosciences, vol 9. Pergamon-Vieweg, Oxford, pp 4–7

Samuelsson B (1982) The leukotrienes: an introduction. In: Samuelsson B, Paoletti R (eds) Advances in prostaglandin, thromboxane and leukotriene research, vol 9. Raven, New York, pp 1–17

Samuelsson B (1983) Leukotrienes: mediators of immediate hypersensitivity reactions and inflammation. Science 220:568–575

Sasaki S, Imai M (1980) Effects of vasopressin on water and NaCl transport across the in vitro perfused medullary thick ascending limb of Henle's loop in mouse, rat, and rabbit kidneys. Pflügers Arch 383:215–221

Sato M, Dunn MJ (1984) Interactions of vasopressin, prostaglandins, and cAMP in rat papillary collecting tubule cells in culture. Am J Physiol 247:F423–F433

Satoh H, Satoh S (1980) Prostaglandin formation by microsomes of dog kidney: prostacyclin is a major prostaglandin of dog renal microsomes. Biochem Biophys Res Commun 94:1266–1272

Satoh S, Zimmermann BG (1975) Influence of the renin-angiotensin system on the effect of prostaglandin synthesis inhibitors on the renal vasculature. Circ Res 37 [Suppl I]:I–89 – I–96

Scherer B, Siess W, Weber PC (1977) radioimmunological and biological measurement of prostaglandins in rabbit urine: decrease of PGE_2 excretion at high NaCl intake. Prostaglandins 13:1127–1139

Scherer B, Schnermann J, Sofroniev M, Weber PC (1978) Prostaglandin (PG) analysis in urine of humans and rats by different radioiummunoassays:-effect on PG-excretion by PG-synthetase inhibitors, laparotomy and furosemide. Prostaglandins 15:225–266

Schlondorff D, Carvounis CP, Jacoby M, Satriano JA, Levine SD (1981) Multiple sites for interaction of prostaglandin and vasopressin in toad urinary bladder. Am J Physiol 241:F625–F631

Schlondorff D, Zanger R, Satriano JA, Folkert VW, Eveloff J (1982) Prostaglandin synthesis by isolated cells from the outer medulla and from the thick ascending loop of Henle of rabbit kidney. J Pharmacol Exp Ther 223:120–124

Schnermann J, Briggs JP (1981) Participation of renal cortical prostaglandin in the regulation of glomerular filtration rate. Kidney Int 19:802–815

Schnermann J, Schubert G, Hermle M, Herbst R, Stowe NT, Varimizi S, Weber PC (1979) The effect of inhibition of prostaglandin synthesis on tubuloglomerular feedback in the rat kidney. Pflugers Arch 379:269–286

Schnermann J, Briggs JP, Weber PC (1984) Tubuloglomerular feedback, prostaglandins, and angiotensin in the autoregulation of glomerular filtration rate. Kidney Int. 25:53–64

Schooley JC, Mahlmann LJ (1971) Stimulation of erythropoiesis in plethoric mice by prostaglandins and its inhibition by antierythropoietin. Proc Soc Exp Biol Med 138:523–524

Schor N, Brenner BM (1981) Possible mechanism of prostaglandin-induced renal vasoconstriction in the rat. Hypertension 3 [Suppl II]:II–81 – II–85

Schor N, Ichikawa I, Brenner BM (1980) Glomerular adaptations to chronic dietary salt restriction or excess. Am J Physiol 238:F428–F436

Schremmer JM, Blank ML, Wykle RL (1979) Bradykinin – stimulated release of [^{3}H] – arachidonic acid from phospholipids of $HSDM_1C_1$ cells: comparison of diacyl phospholipids and plasmalogens as prostaglandin precursors. Prostaglandins 18:491–505

Schryver S, Sanders E, Beierwaltes WH, Romero JC (1984) Cortical distribution of prostaglandin and renin in isolated dog glomeruli. Kidney Int 25:512–518

Schwartz D, DeSchryrer-Kecskemeti K, Needleman P (1984) Renal arachidonic acid metabolism and cellular changes in the rabbit renal vein constricted kidney: inflammation as a common process in renal injury models. Prostaglandins 27:605–613

Schwartzman M, Ferreri NR, Carrol MA, Songu Mize E, McGiff JC (1985) Vasopressin stimulates cytochrome P450-related arachidonic acid metabolism in renal medullary cells. Nature 314:620–622

Secrest RJ, Olsen EJ, Chapnick BM (1985) Leukotriene D_4 relaxes canine renal and superior mesenteric arteries. Circ Res 57:323—329

Seymour AA, Zehr JE (1979) Influence for renal prostaglandin synthesis on renin control mechanisms in the dog. Cir Res 45(I):13—25

Seymour AA, Davis JO, Freeman RH, DeForrest JM, Rowe BP, Williams GM (1979) Renein release from filtering and non-filtering kidneys stimulated by PGI_2 and PGD_2. Am J Physiol 237:F285—F290

Seymour AA, Davis JO, Echtenkamp SF, Dietz JR, Freeman RH (1981) Adrenergically induced renin in conscious indomethacin-treated dogs and rats. Am J Physiol 240(6):F515—F521

Sies W, Siegel FL, Lapetina EG (1984) Dihomogammalinoleic acid but not eicosapentaenoic acid, activates washed human platelets. Biochim Biophys Acta 801:265—276

Sircar JC, Schwender GF, Johnson EA (1983) Soybean lipoxygenase inhibition by non-steroidal anti-inflammatory drugs. Prostaglandins 25:393—396

Smith JB, Willis AL (1971) Aspirin selectively inhibits prostaglandin synthesis in human platelets. Nature 231:235—237

Smith JB, Alam I, Ingerman-Wojenoki CM, Siegl AM, Silver MJ (1982) Prostsglandins as modulators of platelet function. In: Herman AG, Vanhoute PM, Denolin H, Gossens A (eds) Cardiovascular pharmacology of the prostaglandins. Raven, New York

Smith WL, Lands WEM (1972) Oxygenation of polyunsaturated fatty acids during prostaglandin synthesis by sheep vesicular gland. Biochemisty II:3276—3285

Smith WL, Bell TG (1978) Immunohistochemical localization of the prostaglandin-forming cyclooxygenase in renal cortex. Am J Physiol 235:451—457

Smith WL, Wilkin GP (1977) Immunochemistry of prostaglandin endoperoxide-forming cyclooxygenase: the detection of the cyclooxygenases in rat, rabbit and guinea pig kidneys by immunogluorescence. Prostaglandins 13:873—892

Solez K, Fox JA, Miller M, Heptinstall RH (1974) Effects of indomethacin on renal inner medullary plasma flow. Prostaglandins 7:91—98

Spokas EG, Wong PYK, McGiff JC (1982) Prostaglandin-related renin release from rabit renal cortical slices. Hypertension 4 [Suppl II]:II96—II100

Sraer J, Rigaud M, Bens M, Rabinovitch H, Ardaillou R (1983) Metabolism of arachidonic acid via the lipoxygenase pathway in human and murine glomeruli. J Biol Chem 258:4325

Sraer J, Bens M, Ardaillou R, Sraer JD (1986) Sulfidopeptide leukotriene (LT) biosynthesis and metabolism by rat glomeruli (G) and papilla (P). Kidney Int 29(1):346

Stahl RAK, Attallah AA (1980) Sar^1-Ile^{8-} angiotensin II decreases urinary PGE_2 and PGF_{2a} excretion in rabbits on dietary sodium restriction. Eur J Pharmacol 65:81—84

Stahl RAK, Attallah AA, Bloch DL, Lee JB (1979) Stimulation of rabbit renal PGE_2 biosynthesis by dietary sodium restriction. Am J Physiol 237:F344—F349

Stoff JS, Rosa RM, Silva P, Epstein FH (1982) Indomethacin impairs water diruesis in the DI rat: role of prostaglandins independent of ADH. Am J Physiol 241:231—237

Stokes JB (1979) Effect of prostaglandin E_2 on chloride transport across the rabbit thick ascending limb of Henle, selective inhibition of the medullary portion. J Clin Invest 64:495—502

Stokes JB (1981) Integrated actions of renal medullary prostaglandins in the control of water excretion. Am J Physiol 240:F471—F480

Stokes JB, Kokko JP (1977) Inhibition of sodium transport by prostaglandin E_2 across the isolated, perfused rabbit collecting tubule. J Clin Invest 59:1099—1104

Stone KJ, Hart M (1975) Prostaglandin-9-ketoreductase in rabbit kidney. Prostaglandins 10:273—288

Strandhoy JW, Otto CE, Schneider EG, Willis LR, Beck NP, Davis BB, Knox FG (1974) Effects of prostaglandin E_1 and E_2 on renal sodium reabsorption and starling forces. Am J Physiol 226:1015–1021

Susic D, Sparks JC (1975) Effects of aspirin on renal sodium excretion, blood pressure and plasma and extracellular fluid volume in salt-loaded rats. Prostaglandins 10:825–831

Suzuki S, Franco-Saenz R, Ran SY, Mulrow PJ (1981) Effects of indomethacin on plasma renin activity in the conscious rat. Am J Physiol 240:E286–E289

Swain JA, Heyndrickx GR, Borttcher DH, Vatner SF (1975) Prostaglandin control of renal circulation in the unanesthetized dog and baboon. Am J Physiol 229: 826–830

Tan SY, Sandwisch DW, Mulrow PJ (1980) Sodium intake as a determinant of urinary prostaglandin E_2 excretion. Prostaglandins Med 4:53–63

Tannenbaum J, Splawinski JA, Oates JA, Nies AS (1975) Enhanced renal prostaglandin production in the dog. I. Effects on renal function. Circ Res 36:197–203

Terragno NA, Terragno DA, McGiff JC (1977) Contribution of prostaglandins to the renal circulation in conscious, anesthetized and laparatomized dogs. Circ Res 40:590–595

Tobias TD, Vane FM, Paulsrud JR (1975) The biosynthesis of 1a, 1b dihomo-PGE_2 and 1a, 1b-dihomo-$PGF_{2\alpha}$ from 7, 10, 13, 16-docosatetraenoic acid by acetone-pentane powder of sheep vesicular gland microsomes. Prostaglandins 10:443–468

Torikai S, Kurokawa K (1981) Distribution of prostaglandin E_2-sensitive adenylate cyclase along the rat nephron. Prostaglandins 21:427–438

Torikai S, Kurokawa K (1983) Effect of PGE_2 on vasopressin-dependent cell cAMP in isolated single nephron segments. Am J Physiol 245:F58–F66

Torres VE, Strong CG, Romero JC, Wilson DM (1975) Indomethacin enhancement of glycerol-induced acute renal failure in rabbits. Kidney Int 7:170–178

Tyssebotn I, Kirkebo A (1977) The effect of indomethacin on renal blood flow distribution during hemorrhagic hypotension in dog. Acta Physiol Scand 101: 15–21

Ulgesity A, Kreisberg JI, Levine L (1983) Stimulation of arachidonic acid metabolism in rat kidney mesangial cells by bradykinin, antidiuretic hormone, and their analogues. Prostaglandins Leukotrienes Med 10:83–93

Urakabe S, Takamitsu Y, Shirai D, Yuasa S, Kimura G, Orita Y, Abe H (1975) Effect of different prostaglandins on permeability of toad urinary bladder. Comp Biochem Physiol 52:1–4

Usberti M, Dechaux M, Guillot M, et al. (1980) Renal prostaglandin E_2 in nephrogenic diabetes insipidus: effects of inhibition of prostaglandin synthesis by indomethacin. J Pediatr 97:476–478

Valtin H (1976) Animal model of human disease: hereditary hypothalamic diabetes insipidus in the Brattleboro strain of rat. Am J Pathol 83:633–636

Vander AJ (1968) Direct effects of prostaglandin on renal function and renin release in anesthetized dog. Am J Physiol 214:281

Vandongen R, Tunney A, Mahoney D, Barden A (1981) Dissociation of beta-adrenergic stimulation of renin secretion and prostacyclin synthesis in the rabbit kidney. Prostaglandins 21:1007–1014

van Dorp DA, (1966) The biosynthesis of prostaglandins. Mem Soc Endocr 14: 39–47

van Dorp DA, Beerthuis RK, Nugteren DH, Vonkeman H (1964) The biosynthesis of prostaglandins. Biochim Biophys Acta 90:204–207

Vane JR (1971) Inhibition of prostaglandin synthesis as a mechanism of action for aspirin-like drugs. Nature 231:232–235

Vargaftig BB, Dao Hai N (1972) Selective inhibition by mepacrine of the release of "rabbit aorta contracting substance" evoked by the administration of bradykinin. J Pharm Pharmacol 24:159–161

Vatner SF (1974) Effects of hemorrhage on regional blood flow distribution in dogs and primate. J Clin Invest 54:225–235

Vaughn ED, Shenashy JH II, Gillenwater JY (1971) Mechanism of acute hemodynamic response to ureteral occlusion. Invest Urol 9:109–118

Venuto RC, O'Dorisio T, Ferris TF, Stein JH (1975) Prostaglandins and renal function II. The effect of prostaglandin inhibition on autoregulation of blood flow in the intact kidney of the dog. Prostaglandins 9:814–828

Vikse A, Holdaas H, Sejersted OM, Kiil F (1984) Relationship between PGE_2 and renin release in dog kidneys. Effects of efferent artierolar dilation and adrenergic stimulation. Acta Physiol Scand 121:261–268

Villarreal D, Davis JO, Freeman RH, Sweet WD, Dietz JR (1984) Effects of meclofenamate on the renin response to aortic constriction in the rat. Am J Physiol 247:R546–551

von Euler US (1934) Zur Kenntnis der pharmakologischen Wirkungen von Nativsekreten und Extrakten männlicher accessorischer Geschlechtsdrüsen. Naunyn-Schmiedebergs, Arch Exp Pathol Pharmacol, 175:78–84

von Euler US (1935) Kurz wissenschaftliche Mitteilungen. Über die spezifische blutdrucksenkende Substanz des menschlichen Prostata- und Samenblasensekretes. Klin Wochenschr 14:1182–1184

Vonkeman H, van Dorp DA (1968) The action of prostaglandin synthetase on 2-arachidonyllecithin. Biochim Biophys Acta 164:430–432

Walker LA, Frölich JC (1981) Dose-dependent stimulation of renal prostaglandin synthesis by deamino-8-D-arginine vasopressin in rats with hereditary diabetes insipidus. J Pharmacol Exp Ther 217:87–91

Walker LA, Valtin H (1982) Biological importance of nephron heterogeneity. Annu Rev Physiol 44:203–219

Walker LA, Whorton A, Smigel M, France R, Frölich JC (1978) Antiduiretic hormone increase renal prostaglandin synthesis in vivo. Am J Physiol 235:180–185

Walker RM, Stoff JS, Brown RS, Epstein FH (1980) The relation of renal prostaglandins to urinary dilution in lithium-induced nephrogenic diabetes insipidus and normal subjects. Clin Res 28:463

Walker RM, Brown RS, Stoff JS (1981) Role of renal prostaglandins during antidiuresis and water diruesis in man. Kidney Int 21:365–370

Walshe JJ, Venuto RC (1979) Acute oliguric renal failure induced by indomethacin: possible mechanism. Ann Intern Med 91:47–49

Weber P, Holzgreve H, Stephan R, Herbst R (1975) Plasma renin activity and renal sodium and water excretion following infusion of arachidonic acid in rats. Eur J Pharmacol 34:299

Weber PC, Larsson C, Anggard E, Hamberg M, Corey EJ, Nicolaou KC, Samuelsson B (1976) Stimulation of renin release from rabbit renal cortex by arachidonic acid and prostaglandin endoperoxides. Circ Res 39:868–874

Weber PC, Larsson C, Scherer B (1977a) Prostaglandin E_2-9-ketoreductase as a mediator of salt intake-related prostaglandin-renin interaction. Nature 266:65–66

Weber PC, Scherer B, Larsson C (1977b) Increase of free arachidonic acid by furosemide in man as the cause of prostaglandin and renin release. Eur J Pharmacol 41:329–332

Werning C, Vetter W, Weidmann P, Schweikert HU, Stiel D, Siegenthaler W (1981) Effect of prostaglandin E_2 on renin in the dog. Am J Physiol 220:852

Whinnery MA, Shaw JO, Beck N (1982) Thromboxane B_2 and prostaglandin E_2 in the rat kidney with unilateral ureteral obstruction. Am J Physiol 242:F220–F225

Whorton AR, Misono K, Hollifield J, Frölich JC, Inagami T, Oates JA (1977) Prostaglandins and renin release I. Stimulation of renin release from rabbit renal cortical slices by PGI_2. Prostaglandins 14:1095–1104

Whorton AR, Smigel M, Oates JA, Frölich JC (1978) Regional differences in prostaglandin formation by the kidney: Prostacyclin is a major prostaglandin of renal cortex. Biochim Biophys Acta 529:176–180

Whorton AR, Lazar JD, Smigel MD, Oates JA (1980) Prostaglandin mediated renin release from renal cortical slices. In: Samuelsson B, Ramwell PW, Paoletti R (eds) Prostaglandin thromboxane research, vol 7. Raven, New York, pp 1123-1129

Whorton AR, Lazar JD, Smigel MD, Oates JA (1981) Prostaglandins and renin release: III. effects of PGE_1, E_2, F_2, and D_2 on renin release from rabbit renal cortical slices. Prostaglandins 22:455–468

Williams WM, Frölich JC, Nies AS, Oates JA (1977) Urinary prostaglandins: site of entry into renal tubular fluid. Kidney Int 11:256–260

Williamson HE, Bourland WA, Marchand GR (1975) Inhibition of furosemide induced increase in renal blood flow by indomethacin. Proc Soc Exp Biol Med 148: 164–167

Williamson HE, Gaffney GR, Bourland WA, Farley DB, Vanorden DE (1978) Phenylbutazone-induced decreased in renal blood flow. J Pharmacol Exp Ther 204: 130–134

Winokur TS, Morrison AR (1981) Regional synthesis of monohydroxy eicosanoids by the kidney. J Biol Chem 256:10221–23

Wolfe LS, Rostworowski K, Marion J (1976) Endogenous formation of the prostaglandin endoperoxide metabolite, thromboxane B_2, by brain tissue. Biochim Biophys Res Commun 70:907–913

Wong PYK, Malik KU, Desiderio DM, McGiff JC, Sun FF (1980) Hepatic metabolism of prostaglandin I_2 (PGI_2) in the rabbit: Formation of a potent novel inhibitor of platelet aggregation. Biochem Biophys Res Commun 93:486

Work J, Baehler RW, Kotchen TA, Talwalk R, Luke RG (1980) Effect of prostaglandin synthesis inhibition on sodium chloride reabsorption in the diluting segment of the conscious dog. Kidney Int 17:24–30

Wright FS, Schnermann JK (1974) Interference with feedback control of glomerular filtration rate by furosemide, triflocin, and cyanide. J Clin Invest 53:1695–1708

Wright LF, Rosenblatt SG, Lifschitz MD (1981) High urine flow rate increases prostaglandin E excretion in the conscious dog. Prostaglandins 22:21–34

Yarger WE, Schocken DD, Harris RH (1980) Obstructive nephropathy in the rat: possible roles for the renin-angiotensin system, prostaglandins, and thromboxanes in postobstructive renal function. J Clin Invest 64:400–412

Zambraski EJ, Dunn MJ (1979) Renal prostaglandin E_2 secretion and excretion in conscious dogs. Am J Physiol 236:F552–558

Zambraski EJ, Dunn MJ (1984) Importance of renal prostaglandins in control of renal function after chronic ligation of the common bile duct in dogs. J Lab Clin Med 103:549–559

Zenser TV, Davis BB (1977) Effects of prostaglandins on rat and adenylate cyclase-cyclic AMP systems. Prostaglandin 14:437–447

Zenser TV, Herman CA, Gorman RR, Davis BB (1977) Metabolism and action of the prostaglandin endoperoxide PGH_2 in rat kidney. Biochem Biophys Res Commun 79:357–363

Zins GR (1975) Renal prostaglandins. Am J Med 58:14–24

Zipser RD, Martin K (1982) Urinary excretion of arterial blood prostaglandins and thromboxanes in man. Am J Physiol 242:171–177

Zipser RD, Hoefs JC, Speckard PF, Zia PK, Horton R (1979) Prostaglandins. Modulators of renal function and pressor resistance in chronic liver disease. J Clin Endocrinol Metab 48:895–900

Zipser RD, Myers SI, Needleman P (1980) Exaggerated prostaglandin and thromboxane synthesis in the rabbit with renal vein constriction. Circ Res 47:231–237

Zipser RD, Little TE, Wilson W, Duke R (1981a) Dual effects of antidiuretic hormone on urinary prostaglandin E_2 excretion in man. J Clin Endocrinol Metab 53:522—526

Zipser RD, Myers SI, Needleman P (1981b) Stimulation of renal prostaglandin synthesis by the pressor activity of vasopressin. Endocrinology 108:495—499

Zook T, Strandhoy JW (1980) Inhibition of ADH-enhanced transepithelial urea and water movement by prostaglandins. Prostaglandins 20:1—13

Zook TE, Strandhoy JW (1981) Mechanisms of the natriuretic and diuretic effects of PGF_{2a}. J Pharmacol Exp Ther 217:674—682

Zusman RM (1981) Prostaglandin, vasopressin and renal water reabsorption. Med Clin North Am 64:915—925

Zusman RM, Keiser HR (1977a) Prostaglandin biosynthesis by rabbit renomedullary interstitial cells in tissue culture. J Clin Invest 60:215—223

Zusman RM, Keiser HR (1977b) Prostaglandin E_2 biosynthesis by rabbit renomedullary interstitial cells in tissue culture mechanism of stimulation by angiotensin II, bradykinin and arginine vasopressin. J Biol Chem 252:2069—2071

Zusman RM, Spector D, Caldwell BV, Speroff L, Schneider G, Mulrow PJ (1973) The effect of chronic sodium loading and sodium restriction on plasma prostaglandin A, E, and F concentrations in normal humans. J Clin 52:1093—1098

Zusman RM, Snider JJ, Cline A, Caldwell BV, Speroff L (1974) Antihypertensive function of a renal-cell carcinoma: evidence for a prostaglandin-A-secreting tumor. N Engl J Med 290:843—845

Zusman RM, Keiser HR, Handler JF (1977a) Vasopressin-stimulated prostaglandin E biosynthesis in the toad urinary bladder: Effect on water flow. J Clin Invest 60:1339—1347

Zusman RM, Keiser HR, Handler JS (1977b) Inhibition of vasopressin-stimulated prostaglandin E biosynthesis by chlorpropamide in the toad urinary bladder. J Clin Invest 60:1348—1353

Zusman RM, Keiser HR, Handler JS (1978) Effect of adrenal steroids on vasopressin-stimulated PGE synthesis and water flow. Am J Physiol 234:532—540

Rev. Physiol. Biochem. Pharmacol., Vol. 107
© by Springer-Verlag 1987

Presynaptic α-Autoreceptors

KLAUS STARKE

Contents

Pharmakologisches Institut, Hermann-Herder-Straße 5, D-7800 Freiburg i. Br., Federal Republic of Germany
The article developed from the author's Rudolf-Buchheim-Lecture at Aachen, September 1985. Work in the author's laboratory was supported by the Deutsche Forschungsgemeinschaft (SFB 70 and AFB 325)

Abbreviations and Code Numbers

AH 21-132, cis-6-(p-acetamidophenyl)-1,2,3,4,4a,10b-hexahydro-8,9-
    dimethoxy-2-methyl-benzo[c] [1,6]-naphthyridine
4-AP, 4-aminopyridine
Bay K 8644, methyl 1,4-dihydro-2,6-dimethyl-3-nitro-4-(2-trifluoro-
    methylphenyl)-pyridine-5-carboxylate
B-HT 920, 6-allyl-2-amino-5,6,7,8-tetrahydro-4H-thiazolo-[4,5-d]-
    azepine
EGTA, ethyleneglycol-bis-(β-aminoethyl ether)-N,N,N',N'-tetraacetic
    acid
IAP, islet-activating protein
IBMX, 3-isobutyl-1-methylxanthine
ICI 63,197, 2-amino-6-methyl-5-oxo-4-n-propyl-4,5-dihydro-s-triazolo
    [1,5-a]pyrimidine
LSD, lysergic acid diethylamide
α-MT, α-methyl-p-tyrosine
5-OCH$_3$-T, 5-methoxy-tryptamine
TEA, tetraethylammonium

1 Introduction

This essay reviews current *quaestiones disputatae* on the presynaptic
α-autoreceptors of noradrenergic neurons. Many more neurons — such as
adrenergic, dopaminergic, histaminergic, serotonergic, cholinergic, and
γ-aminobutyric acid-containing neurons — are now thought to possess
autoreceptors. Yet the α-autoreceptors have certainly aroused the greatest
interest — interest for their own sake, interest because of their possible
clinical implications, and interest, of course, because their study initiated
the α_1/α_2 subclassification. Moreover, methodological approaches to,
and patterns of thought on, the other autoreceptors mirror approaches
to and ideas on presynaptic α-autoreceptors.

Presynaptic autoreceptors are receptors located on or close to
the axon terminals of a neuron, through which the neuron's own transmitter can and, under appropriate conditions, does modify transmitter
biosynthesis or depolarization-evoked transmitter release (see Starke and
Langer 1979). Noradrenergic terminal axons may possess at least two
kinds of autoreceptor: α-adrenoceptors (release-inhibiting, first postulated
in 1971) and β-adrenoceptors (release-enhancing, first postulated in
1975). The β-adrenoceptors are activated only negligibly by released
noradrenaline and, hence, do not fully satisfy the autoreceptor defini-

tion (Langer 1981; Majewski 1983). Presynaptic α-adrenoceptors, in contrast, are activated by released noradrenaline under many conditions. Noradrenergic neurons possess α-adrenoceptors also at their cell bodies and dendrites, and activation of these soma-dendritic receptors leads to a decrease in firing rate. At least in principle, two important functions for which noradrenergic (and other) neurons were devised seem to be subject to modulation by autoreceptors: the generation of action potentials and the release of transmitter by action potentials.

I shall pick up in this essay the thread of previous review articles (Langer 1977, 1981; Starke 1977, 1981a, b; Westfall 1977; Vizi 1979; Gillespie 1980; Rand et al. 1980). As in those articles, the discussion will be based mainly on studies which were carried out on isolated tissues and in which the release of noradrenaline was estimated from the overflow of noradrenaline (or radiolabeled compounds after treatment with radiolabeled noradrenaline) into the incubation or perfusion fluid. However, studies in which the effector cell response was used to measure transmitter release will also be considered where appropriate.

2 Modulation by Agonists and Antagonists

As the α_1/α_2 classification developed, the presynaptic α-autoreceptors of some tissues (rabbit heart, cat spleen, rabbit pulmonary artery) became the prototype α_2-receptors, and it still seems that presynaptic α-autoreceptors are at least mainly α_2. The autoreceptor hypothesis, then, predicts that agonists at α_2-adrenoceptors such as clonidine and noradrenaline itself should depress depolarization-evoked release of noradrenaline, whereas antagonists such as phenoxybenzamine, phentolamine, yohimbine, and idazoxan should enhance it. Many investigators have been impressed by the regularity with which these effects occur. Are they universal?

Lists showing effects of α-receptor agonists and antagonists on the release of noradrenaline have been compiled previously (Starke 1977; Westfall 1977; Gillespie 1980) and are continued here (Tables 1 and 2). These tables are an attempt at an unbiased summary (criteria: isolated tissues, electrical stimulation, overflow methods); they are not a selection of data that confirm the autoreceptor hypothesis. The previous and present lists jointly indicate that inhibition of release by α_2-receptor agonists and enhancement of release by α_2-antagonists have been observed in the heart of the cat, rat, mouse, guinea pig, and rabbit; a variety of blood vessels from dog, cat, rat, guinea pig, rabbit, cattle, and man; the spleen of the dog, cat, and calf; the kidney of the dog, rat,

Table 1. Effect of α-adrenoceptor agonists on electrically evoked release of noradrenaline

Tissue	Drug and release-inhibiting concentration in μmol/liter	Comment[a]	References
		Dog	
Aorta, mesenteric artery, splenic artery and capsule, portal vein	Noradrenaline 1.2	2 Hz. Cocaine	Lorenz et al. 1979
Basilar, mesenteric, and renal artery	Clonidine 0.0001 − 1	5 Hz. No effect of phenylephrine 0.0001 − 0.01	Sakakibara et al. 1982
Saphenous vein	Clonidine 1	Endogenous noradrenaline. 1 − 5 Hz. Antagonism by phentolamine	Saelens and Williams 1983
Saphenous vein	Noradrenaline 0.1 − 1.2 Adrenaline 0.1 − 1 α-Methylnoradrenaline 0.03 − 0.1 Methoxamine 30 Clonidine 0.01 − 1	1 − 2 Hz. Cocaine, corticosterone, propranolol, indomethacin. Inhibition by clonidine only in absence of cocaine and corticosterone. No effect of phenylephrine 0.1 − 10. Antagonism by phenoxybenzamine, phentolamine	Lorenz et al. 1979; Sullivan and Drew 1980
		Cat	
Middle cerebral artery	Phenylephrine 1 − 10 Clonidine 0.01 − 10 Oxymetazoline 0.01 −10	2 Hz. Cocaine, corticosterone, propranolol. Antagonism by rauwolscine	Skärby 1984
		Rat	
Heart	Noradrenaline 0.005 − 0.5 α-Methylnoradrenaline 0.013 − 4 Clonidine 0.01 − 10 Oxymetazoline 0.005 − 10 Xylazine 0.05 − 5	Perfused heart. 0.01 − 3 Hz. Cocaine, desipramine, corticosterone, propranolol. No effect of methoxamine 0.05 − 100. Antagonism by phenoxybenzamine, phentolamine, yohimbine	Fuder et al. 1983, 1984, 1986

Table 1 (continued)

Tissue	Drug and release-inhibiting concentration in μmol/liter	Comment[a]	References
Heart	Noradrenaline 1 Adrenaline 1 Methoxamine 10 Clonidine 1	Atria. 0.5 (methoxamine, clonidine) or 2 Hz. Cocaine. Antagonism by phenoxybenzamine or (methoxamine) prazosin	Majewski et al. 1981; Medgett and Rand 1981; Story et al. 1985
Tail artery	Methoxamine 3 Clonidine 1	Perfused segments. 1 Hz. Cocaine, propranolol, indomethacin. Antagonism of prazosin against methoxamine and of idazoxan against clonidine	Medgett and Rand 1981; Hicks et al. 1986
Mesenteric artery	Clonidine 0.1	Perfused vasculature. 2 Hz	Su and Kubo 1984
Vena cava	Noradrenaline 0.01 −1 Adrenaline 0.1 − 1	2 Hz. Desipramine, corticosterone. Antagonism by rauwolscine	Göthert and Kollecker 1986
Portal vein	Noradrenaline 0.03 − 1	2 Hz. Cocaine, desipramine, corticosterone. Antagonism by yohimbine	Török et al. 1985; Pernow et al. 1986
Kidney	Noradrenaline 0.01 − 0.1 Adrenaline 0.01 − 0.1 Methoxamine 10 Clonidine 0.1	Perfused kidney. 0.5 − 1 Hz. Cocaine. Antagonism by phentolamine, prazosin (against methoxamine) and idazoxan (against clonidine)	Steenberg et al. 1983; Lokhandwala and Steenberg 1984; Rump and Majewski 1987
Anococcygeus muscle	Clonidine 0.003 − 1 Oxymetazoline 0.003 − 1 Naphazoline 0.01 − 1	2 Hz. Desipramine, metanephrine. No effect of phenylephrine 0.1 − 1 and methoxamine 1. Antagonism by yohimbine	Leighton et al. 1979
Submandibular gland	Noradrenaline 0.4	Perfused gland. 1 − 100 Hz. Desipramine, corticosterone, atropine	Wakade and Wakade 1984

Table 1 (continued)

Tissue	Drug and release-inhibiting concentration in μmol/liter	Comment[a]	References
Hypothala-mus	Adrenaline 0.1 B-HT 920 0.1	Endogenous noradren-aline. 5 Hz. Cocaine. Antagonism by yohimbine	Ueda et al. 1983; Kubo et al. 1986
Nuclei ant. hypothalami and tractus solitarii	Clonidine 0.1 − 10	0.3 − 3 Hz. Desipramine	Cichini et al. 1986
		Guinea pig	
Ileum	Noradrenaline 0.3	1.5 Hz. Desipramine, normetanephrine. Antagonism by yohimbine	Alberts and Stjärne 1982
Gall bladder	Clonidine 50	5 Hz. Atropine	Doggrell and Vincent 1980
		Rabbit	
Aorta	Noradrenaline 0.01 − 0.3 Phenylephrine 1 Clonidine 0.01 − 10 Xylazine 1 − 10	1 − 2 Hz. Cocaine, cor-ticosterone, hydro-cortisone, propranolol	Docherty and Starke 1982; Henseling 1983
Basilar artery	Clonidine 1	Endogenous nor-adrenaline. 2 Hz. Co-caine, desoxy-corticosterone	Duckles 1982
Portal vein	Phenylephrine 1 Clonidine 0.01 − 10 Xylazine 1 − 10	2 Hz. Cocaine, cortico-sterone, propranolol	Docherty and Starke 1982
Kidney	Clonidine 0.01 Phenylephrine 1	Perfused kidney. Endo-genous noradrenaline. 5 − 10 Hz. Antagonism by yohimbine and acetylsalicylic acid	Ercan 1983
Cerebral cortex	Noradrenaline 0.01 − 1 α-Methylnoradrenaline 0.01 − 1 Clonidine 0.01 − 1 Oxymetazoline 0.01 − 1 Xylazine 0.1 − 10	2 Hz. Cocaine. No effect of phenyl-ephrine 0.1 − 1	Reichenbacher et al. 1982

Table 1 (continued)

Tissue	Drug and release-inhibiting concentration in μmol/liter	Comment[a]	References
Hippocampus	Phenylephrine 1 Clonidine 0.01 − 1	3 Hz. Cocaine. Antagonism by yohimbine; no antagonism of prazosin against phenylephrine	Jackisch et al. 1984
Hypothalamus	Noradrenaline 0.1 − 1 Adrenaline 0.01 − 1 Clonidine 1	3 − 5 Hz. Cocaine. Antagonism by idazoxan, yohimbine	Galzin et al. 1982, 1984

Man

Tissue	Drug and release-inhibiting concentration in μmol/liter	Comment[a]	References
Pulmonary artery	Noradrenaline 0.01 − 1 Adrenaline 0.001 − 0.1 α-Methylnoradrenaline 0.1 − 1 Clonidine 0.01 − 10 B-HT 920 0.01 − 1	2 Hz. Cocaine, corticosterone, propranolol. No effect of methoxamine 0.01 − 1. Antagonism by rauwolscine	Hentrich et al. 1986
Saphenous vein	Noradrenaline 0.01 − 1 α-Methylnoradrenaline 0.1 − 1 Clonidine 1 − 10 B-HT 920 0.1 − 10	2 − 6 Hz. Cocaine, corticosterone, propranolol. No effect of methoxamine 0.1 − 10. Antagonism by rauwolscine	Janssens and Verhaeghe 1983; Göthert et al. 1984
Hand and foot blood vessels	Clonidine 0.1	Perfused segments. 2 Hz	Stevens and Moulds 1982
Vas deferens	Clonidine 0.01 − 10	4 − 5 Hz. No effect of phenylephrine 0.001 − 1. Antagonism by phentolamine	Belis et al. 1982; Hedlund et al. 1985

Only tissues not yet registered in the tables of Starke (1977), Westfall (1977), or Gillespie (1980) or registered there on the basis of postsynaptic response measurements or of preliminary evidence only are entered, and only studies on isolated tissues in which release was determined as overflow.

[a] Unless stated otherwise, tissues were incubated or superfused, and the electrically evoked overflow was determined with the aid of ^{3}H-noradrenaline. Frequency, auxiliary drugs (not necessarily present in all experiments), and antagonist effects (not necessarily against all agonists) are indicated.

Table 2. Effect of α-adrenoceptor antagonists on electrically evoked release of noradrenaline

Tissue	Drug and release-enhancing concentration in μmol/liter	Comment[a]	References
		Dog	
Coronary arteries	Phentolamine 1	2 Hz. No effect of prazosin 0.05	Cohen et al. 1983
Basilar, mesenteric, and renal artery	Yohimbine 0.001 − 1 Corynanthine 0.1 − 10	5 Hz	Sakakibara et al. 1982
Pulmonary artery	Yohimbine 0.01 − 1	2 Hz. No effect of prazosin 0.001 − 0.1	Constantine et al. 1980
Saphenous vein	Phentolamine 1	Endogenous nor-adrenaline. 1 − 5 Hz	Saelens and Williams 1983
Saphenous vein	Phenoxybenzamine 3 Phentolamine 0.1 − 10 Yohimbine 0.01 − 1 Prazosin 1	1 − 5 Hz. Cocaine, corticosterone, hydrocortisone, propranolol, indomethacin	Guimarães et al. 1978; Lorenz et al. 1979; Sullivan and Drew 1980
Vas deferens	Phentolamine 1	4 Hz	Belis et al. 1982
		Cat	
Middle cerebral artery	Rauwolscine 0.001 −10 Prazosin 0.1 − 1	2 Hz. Cocaine, corticosterone, propranolol	Skärby 1984
		Rat	
Heart	Yohimbine 1	Perfused heart. Endogenous noradrenaline. 1 − 4 Hz	Dart et al. 1984
Heart	Phentolamine 1 Yohimbine 0.1 − 1	Perfused heart. 3 Hz. Cocaine, propranolol, indomethacin	Fuder et al. 1984
Heart	Phenoxybenzamine 10 Phentolamine 0.1 − 3 Idazoxan 10 Yohimbine 0.1 − 1 Prazosin 0.1 − 3	Atria or ventricle strips. 2 − 10 Hz	Majewski et al. 1981; Medgett and Rand 1981; Bradley and Doggrell 1983; Enero 1984; Loiacono et al. 1985; Story et al. 1985

Table 2 (continued)

Tissue	Drug and release-enhancing concentration in μmol/liter	Comment [a]	References
Tail artery	Yohimbine 1	Endogenous noradrenaline. 10 Hz	Westfall et al. 1985
Tail artery	Phentolamine 0.03 − 3 Idazoxan 1 Yohimbine 0.1 Corynanthine 1 Prazosin 0.1 − 0.3	Perfused segments. 0.3 − 5 Hz. Cocaine, propranolol, indomethacin	Medgett and Rand 1981; Hicks et al. 1986
Vena cava	Phentolamine 0.1 Rauwolscine 0.01 − 1	2 Hz. Desipramine, corticosterone	Göthert and Kollecker 1986
Kidney	Phentolamine 0.053 − 1 Idazoxan 0.1 Corynanthine 0.3 Prazosin 0.1	Perfused kidney. 0.5 − 2 Hz. Cocaine	Steenberg et al. 1983; Lokhandwala and Steenberg 1984; Rump and Majewski 1987
Anococcygeus muscle	Phenoxybenzamine 0.05 − 1 Phentolamine 0.01 − 1 Yohimbine 0.01 − 1	2 Hz. Desipramine, metanephrine	Leighton et al. 1979
Hypo- thalamus	Yohimbine 0.01 −1 Prazosin 1	Endogenous noradrenaline. 5 Hz. Cocaine	Ueda et al. 1983
Hypo- thalamus	Idazoxan 1 Yohimbine 1	3 Hz	Galzin et al. 1984
		Guinea pig	
Portal vein	Phentolamine 1	0.5 − 10 Hz	Muramatsu et al. 1980
Gall bladder	Phentolamine 1	5 Hz. Atropine	Doggrell and Vincent 1980
		Rabbit	
Aorta	Yohimbine 0.01 − 10 Rauwolscine 0.01 − 1	2 − 8 Hz. Cocaine, corticosterone, pro- pranolol. No effect of prazosin 0.001 − 0.1	Docherty and Starke 1982
Basilar artery	Phenoxybenzamine 10 Phentolamine 1 Yohimbine 10	Endogenous nor- adrenaline. 2 − 8 Hz. Cocaine, desoxycorticosterone	Duckles 1982

Table 2 (continued)

Tissue	Drug and release-enhancing concentration in μmol/liter	Comment [a]	References
Mesenteric artery	Phenoxybenzamine 0.1 − 1 Phentolamine 0.1 − 10 Yohimbine 0.1 − 1 Prazosin 1	Endogenous noradrenaline. 10 Hz	Mishima et al. 1984
Ileocolic artery	Phenoxybenzamine 1 Phentolamine 3 Yohimbine 3	Perfused segments. 5 Hz. No effect of prazosin 0.3	von Kügelgen and Starke 1985
Kidney	Yohimbine 0.1 − 1	Perfused kidney. 3 Hz	Hedqvist 1981
Cerebral cortex	Phentolamine 0.1 − 10 Tolazoline 0.1 − 10 Idazoxan 0.1 − 10 Yohimbine 0.01 − 1 Rauwolscine 0.003 − 1 Corynanthine 0.1 − 10	0.3 − 3 Hz. Cocaine. No effect of prazosin 0.1	Reichenbacher et al. 1982; Heepe and Starke 1985
Hippocampus	Yohimbine 0.01 − 10	3 Hz. Cocaine. No effect of prazosin 0.1	Jackisch et al. 1984
Hypo-thalamus	Phentolamine 1 − 10 Idazoxan 0.1 − 100 Yohimbine 0.1 − 1	3 − 5 Hz. Cocaine	Galzin et al. 1982, 1984
		Cattle	
Splenic vein	Phentolamine 1	1 Hz. Cocaine, normetanephrine	Dzielak et al. 1983
		Man	
Pulmonary artery	Rauwolscine 0.01 − 1	2 Hz. Cocaine, cortico-sterone, propranolol. No effect of prazosin 0.01 − 1	Hentrich et al. 1986
Saphenous vein	Yohimbine 0.01 − 1 Rauwolscine 0.01 − 3	2 − 6 Hz. Cocaine, corticosterone, propranolol. No effect of prazosin 0.01 − 1	Janssens and Verhaeghe 1983; Göthert et al. 1984; Docherty and Hyland 1985
Hand and foot blood vessels	Phenoxybenzamine 1 − 10 Phentolamine 1 − 10 Yohimbine 0.01 − 0.1 Rauwolscine 0.001 − 0.1	Perfused segments. 2 Hz. No effect of prazosin 0.01 − 0.1	Stevens and Moulds 1982, 1985

Table 2 (continued)

Tissue	Drug and release-enhancing concentration in μmol/liter	Comment [a]	References
Vas deferens	Rauwolscine 0.01 − 1	4 − 8 Hz. No effect of phentolamine 1 − 10, yohimbine 1, and prazosin 0.001 − 0.1	Belis et al. 1982; Hedlund et al. 1985

Only tissues not yet registered in the tables of Starke (1977) and Gillespie (1980) or registered there on the basis of postsynaptic response measurements or of preliminary evidence only are entered, and only studies on isolated tissues in which release was determined as overflow.

[a] Unless stated otherwise, tissues were incubated or superfused, and the electrically evoked overflow was determined with the aid of ^{3}H-noradrenaline. Frequency and auxiliary drugs (not necessarily present in all experiments) are indicated.

and rabbit; the rat submandibular gland; the intestine of the cat and guinea pig; the guinea-pig gall bladder; the rat anococcygeus muscle; the rat iris; the nictitating membrane of the cat; the vas deferens of dog, rat, mouse, guinea pig, rabbit, and man; the human oviduct; the superior cervical ganglion of the rat and rabbit; and various brain regions of the rat, guinea pig and rabbit. This catalog can be extended further when we add isolated tissues in which noradrenaline was released by high K^+ concentrations, and organs in which the venous overflow of noradrenaline was measured in vivo. They include the dog heart (e.g., Saeed et al. 1985; see also Sect. 6.3), the pig spleen (Lundberg et al. 1986), the pineal gland of the rat (Pelayo et al. 1978), the rat hippocampus (Frankhuyzen and Mulder 1982) and thalamus (Bradberry and Adams 1986) and the hypothalamus of the possum (Minson and de la Lande 1984).

Are there, then, no negative findings? Hardly any; yet some reservations must be expressed.

1. A clearly negative finding was obtained in the median eminence of the rat. Neither xylazine nor yohimbine modified the release of noradrenaline, thus suggesting the absence of $α_2$-autoreceptors (Vizi et al. 1985).

2. Phenoxybenzamine has occasionally been used at high concentrations that may have inhibited the reuptake of noradrenaline. This, and not blockade of presynaptic receptors, may for instance have increased the evoked overflow of noradrenaline from the dog kidney in the study by Zimmerman et al. (1971). In a later postsynaptic response study, no evidence for presynaptic α-adrenoceptors in dog kidney was found (Robie 1980).

3. Results obtained in the rabbit kidney are also equivocal. The release-enhancing effect of yohimbine persisted in the presence of indomethacin (Table 2). Yet the inhibitory effect of clonidine was prevented by another cyclooxygenase inhibitor, namely, acetylsali-cyclic acid, and hence seemed to be mediated by prostaglandins (Table 1).

4. Inhibitory agonist and facilitatory antagonist effects were observed in rat spleen (Kalsner and Quillan 1984), guinea-pig ureter (Kalsner 1982; Kalsner and Quillan 1984), and cattle iris (Kalsner 1983). These tissues were not entered in Tables 1 and 2 because the authors suggest that at least the antagonists do *not* act through presynaptic α-adreno-ceptors.

5. The effects of α-adrenoceptor antagonists in cattle arteries were variable (Kalsner and Chan 1979). Are more "side effects" superimposed on the typical release-enhancing α-antagonist effect in cattle arteries than in other tissues? Such side effects might include release by phento-lamine of histamine (which then might depress the release of nor-adrenaline; Dzielak et al. 1983) and, in the case of yohimbine, local anesthesia (Hagan and Hughes 1986) and activation of presynaptic inhibitory serotonin receptors (Feuerstein et al. 1985).

Like other presynaptic modulators, α_2-adrenoceptor ligands may not act in the same way on all noradrenergic axons, across the tissues and species. Notwithstanding the reservations just expressed, however, it seems safe to conclude that under appropriate conditions (see Sect. 6.1) all (and not only certain) α_2-agonists reduce, and all (and not only a few) α_2-antagonists increase, the release of noradrenaline in the over-whelming majority of tissues.

3 The Common Receptor

α_2-Receptor agonists and antagonists act at one and the same receptor in many cells such as cholinergic neurons, platelets, and myocytes, and compete with one another for specific α_2-radioligand binding sites. All this appears to leave little doubt that they produce their effects on noradrenergic terminal axons equally at the same receptor, a receptor, moreover, at which endogenous noradrenaline also causes inhibition. The α_2-autoreceptor as a target of released noradrenaline will be discussed in more detail later (see Sect. 6.1). This chapter summarizes direct evidence for a common site of action of exogenous α_2-receptor agonists and antagonists. Subsequently, indications will be discussed that pre-synaptic α-adrenoceptors may not be entirely homogeneous, firstly, because presynaptic α_1-receptors may also occur, and secondly, because there may be differences within the group of presynaptic α_2-receptors itself.

3.1 Interaction Between Agonists and Antagonists

Four kinds of interaction experiment support a common site for exogenous α_2-receptor agonists and antagonists. The findings must be evaluated with care, however, because, other than in simpler systems, a third compound — released noradrenaline — may also compete for this site.

1. Antagonists, when given at relatively high concentrations, prevent the effect of low concentrations of agonists. Yet the antagonists themselves will increase the release of noradrenaline in most experiments of this kind, and then one is faced with an interpretational difficulty: the reduction in agonist effect may be due to the presence of the antagonist (genuine antagonism), but also to the increase in release *per se* (because any increase in the biophase concentration of noradrenaline, independently of its cause, must narrow the scope for exogenous α-adrenergic inhibition, and because procedures that reinforce the releasing stimulus generally diminish presynaptic modulation; see Sect. 5.2). In most cases of antagonism listed in Table 1, these alternatives have not been distinguished. Evidence for genuine antagonism can be obtained, however, when the antagonist-induced facilitation is either compensated for or avoided. A compensation was attempted in the first two studies that showed a release-inhibiting effect of exogenous noradrenaline itself. Phenoxybenzamine and phentolamine counteracted the effect of noradrenaline even when the frequency of stimulation was reduced and, hence, release was low despite the antagonist: the presence of the antagonist, and not the rise in transmitter release *per se,* rendered the exogenous amine ineffective (Starke 1972b; Kirpekar et al. 1973). An enhancement of release can be avoided in superfused synaptosomes, since any released noradrenaline is rapidly washed away by the superfusion stream and, hence, autoinhibition does not develop. In rat and guinea-pig cerebrocortical synaptosomes, phentolamine and yohimbine had no facilitatory effect of their own and yet prevented the inhibitory effect of exogenous noradrenaline and clonidine (de Langen et al. 1979; Ebstein et al. 1982; Raiteri et al. 1983, 1984). These examples of antagonism, undistorted by any effect of the antagonist alone, speak convincingly for a common site.

2. Agonists, when given at relatively high concentrations, conversely prevent the effect of relatively low concentrations of antagonists

(see Starke 1977 and, for a recent example, Göthert et al. 1984). Additional findings analogous to those discussed in the preceding paragraph indicate that this is due to the presence of the high agonist concentration and not to the ensuing agonist-induced decrease in transmitter release *per se.*

3. α_2-Adrenoceptor antagonists shift agonist concentration-response curves to the right. In simpler systems, such shifts provide strong support for competition, in particular when the slope of the Schild regression line is unity, and moreover allow the estimation of antagonist affinities. In an investigation of autoreceptors the interpretation is again more difficult. Table 3 summarizes studies carried out with overflow methods. From the pA_2 values it can be seen that the phentolamine dissociation constants vary 66-fold, the yohimbine constants 21-fold, and the rauwolscine constants even 275-fold. Part of this variation is certainly due to the presence of varying concentrations of the third rival — released noradrenaline — in the biophase, and it has been emphasized that the low pA_2 values of phentolamine in the rabbit pulmonary artery are underestimates of the "true" value for that reason (Starke et al. 1974, 1975a). In only one study in Table 3 were the conditions of nerve stimulation chosen such as to keep the biophase concentration of noradrenaline too low for autoinhibition, so that the antagonists did not cause an increase and a clean two-compound system was obtained (Fuder et al. 1983). The shifts to the right of the oxymetazoline concentration-response curve that phentolamine and yohimbine caused under these conditions, leading to Schild plot slopes near unity, are particularly clear demonstrations of competition for one presynaptic receptor (see also analogous postsynaptic response studies, for instance in rat vas deferens: Drew 1977; Tayo 1979; Leedham and Pennefather 1982; Warming et al. 1982; Doxey et al. 1984; Lattimer and Rhodes 1985 — all with Schild plots).

4. Reversibly acting α-agonists and -antagonists protect noradrenergic axons against the long-lasting and perhaps irreversible release-enhancing effect of phenoxybenzamine. Early experiments of this kind were carried out by McCulloch et al. (1972) and Farah and Langer (1974). A third receptor protection experiment by Kalsner and Chan (1979) gave a negative result; however, the failure of phentolamine (30 μmol/liter) to prevent the release-enhancing effect of phenoxybenzamine (30 μmol/liter) in cattle facial arteries may have been due to the high phenoxybenzamine:phentolamine concentration ratio (1:1 instead of the 0.027:1 used by Farah and Langer 1974). We recently tried to protect the noradrenergic axons of rabbit brain cortex slices against phenoxybenzamine by both agonists and antagonists. As shown in Fig. 1,

Table 3. pA$_2$ values of α-adrenoceptor antagonists at presynaptic α$_2$-autoreceptors

Tissue	Agonist	pA$_2$ value of			References
		Phentolamine	Yohimbine	Rauwolscine	
Cat middle cerebral artery	Clonidine			9.7 [a]	Skärby 1984 [b]
Rat heart	Oxymetazoline	7.52 [c]	7.82 [c]		Fuder et al. 1983
Rat vena cava	Noradrenaline			7.58	Göthert and Kollecker 1986
Rat kidney	Noradrenaline	8.0 [a]			Lokhandwala and Steenberg 1984
Rat cerebral cortex	Noradrenaline Clonidine	8.0 [c]	7.4 [c] 7.7 [c]		Werner et al. 1979 Ennis and Lattimer 1984
Rat hippocampus	?	7.88	6.70		Frankhuyzen and Mulder 1982 [b]
Guinea-pig ileum	Noradrenaline		6.5		Alberts and Stärne 1982
Rabbit pulmonary artery	Noradrenaline Clonidine Oxymetazoline	6.18 6.27 6.30			Starke et al. 1974, 1975a
Rabbit hippocampus	Clonidine		7.7 [a]		Jackisch et al. 1984
Rabbit hypothalamus	Adrenaline		7.5 [a]		Galzin et al. 1982
Human pulmonary artery	B-HT 920			7.26	Hentrich et al. 1986
Human saphenous vein	B-HT 920			7.57	Göthert et al. 1984

The table is based on studies on isolated tissues in which depolarization-evoked release of noradrenaline was measured by overflow. [a] Estimated from authors' data. [b] Prazosin had little or no antagonist effect. [c] Determined from Schild plots; slopes were close to unity

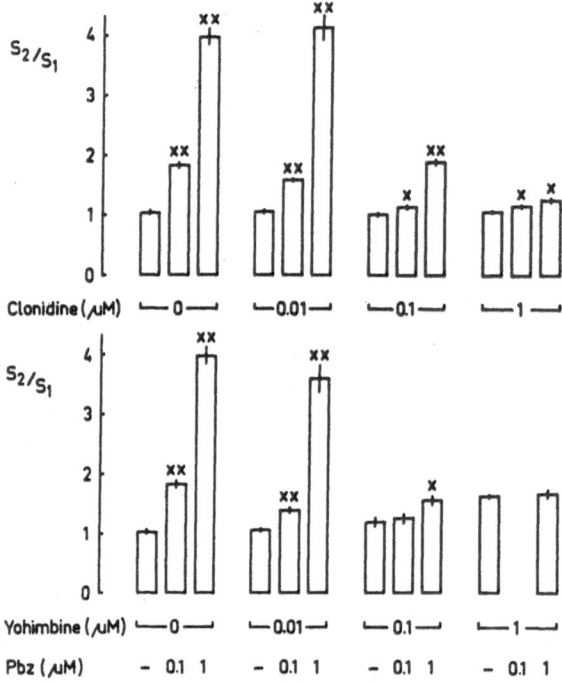

Fig. 1. Interaction of clonidine and yohimbine with phenoxybenzamine in rabbit cerebrocortical slices preincubated with ^{3}H-noradrenaline. After preincubation, the slices were superfused with ^{3}H-noradrenaline-free medium. They were stimulated electrically twice for 2 min each at 3 Hz, after 60 and 250 min of superfusion (S_1, S_2). Phenoxybenzamine (*Pbz*) was added, at concentrations indicated, from 85 to 95 min of superfusion; clonidine or yohimbine was added, at concentrations indicated, from 80 to 95 min superfusion and then washed out together with phenoxybenzamine. *Columns* show ratios between the overflow of tritium evoked by S_2 and S_1 (S_2/S_1). $\times$ and $\times\times$ indicate significant differences (P $<$ 0.05 and 0.01, respectively) from corresponding experiments without phenoxybenzamine. From Hölting and Starke (1986)

release of previously incorporated ^{3}H-noradrenaline was elicited by two periods of electrical stimulation (S_1, S_2). Superfusion with phenoxybenzamine (0.1 and 1 μmol/liter) for 10 min between S_1 and S_2, followed by 155 min of washout, increased the S_2/S_1 ratio 1.8- and 4-fold, respectively (left-hand groups of three columns). Addition of clonidine (0.01 μmol/liter) or yohimbine (0.01 μmol/liter) during the exposure to phenoxybenzamine caused little change. Higher concentrations of clonidine and yohimbine, however, afforded marked protection, thus indicating that all three, i.e., the antagonists phenoxybenzamine and yohimbine and the agonist clonidine, and by inference also endogenous noradrenaline, modulated the release of noradrenaline through the same site.

These four groups of observations indicate that exogenous α_2-receptor agonists and antagonists, by and large, share a common receptor in the terminal region of noradrenergic neurons, namely the presynaptic α_2-autoreceptor. α_2-Agonists and -antagonists also share a common receptor in the soma-dendritic region of the same neurons, namely, the soma-dendritic α_2-receptor (Brown and Caulfield 1979; McAfee et al. 1981; Marwaha and Aghajanian 1982; Williams et al. 1985). Presynaptic and soma-dendritic findings reinforce each other.

3.2 Presynaptic α_1-Adrenoceptors?

The existence of presynaptic α_1-receptors in addition to α_2-receptors was suggested by studies on the heart of dogs (Cavero et al. 1979; Uchida et al. 1984), rats (Kobinger and Pichler 1980; Bradley and Doggrell 1983; Docherty 1984; Story et al. 1985), and guinea pigs (Ledda and Mantelli 1984), and on the rat vas deferens (Docherty 1984; Warnock et al. 1985), tail artery (Hicks et al. 1986), and kidney (Rump and Majewski 1987). In most of these experiments, postsynaptic responses were determined to quantify transmitter release, and postsynaptic effects of the α_1-adrenoceptor ligands used have not always been ruled out (see the criticism by de Jonge et al. 1986). An examination of Tables 1 and 2 shows that the evidence from overflow experiments is not very favourable for presynaptic α_1-receptors; for instance, the α_1-selective compounds phenylephrine and, with four exceptions, prazosin did not change the release of noradrenaline at concentrations below 1 μmol/liter. Positive evidence comes mostly from studies on rat heart, tail artery, and kidney. In these tissues, prazosin at 0.1 μmol/liter and higher concentrations increased the evoked overflow of tritium (after pretreatment with ^{3}H-noradrenaline), the α_1-selective agonist methoxamine at 3–10 μmol/liter caused a decrease, and the effect of methoxamine was antagonized by prazosin (heart: Bradley and Doggrell 1983; Story et al. 1985; tail artery: Hicks et al. 1986; kidney: Rump and Majewski 1987). However, reasons for skepticism remain. Very high concentrations of methoxamine and prazosin were required; for instance, prazosin 0.1 μmol/liter is more than 100 times its α_1-receptor dissociation constant as determined by radioligand binding (see Table 6 of Starke 1981a). In several studies prazosin (at as little as 0.01 μmol/liter in the report by Bradley and Doggrell 1983) accelerated the basal efflux of tritium. This is a known effect of prazosin; it results presumably from a reserpine-like action on noradrenaline storage granules, and such an action can by itself augment action potential-evoked noradrenaline release (Cubeddu and Weiner 1975b), independently of α-adrenoceptor antagonism. It should also be noted that a release-inhibiting effect of methoxamine in rat heart has not been found by all investigators (Fuder et al. 1984). Wetzel et al. (1985) have recently shown that noradrenaline inhibits the release of ^3H-*acetylcholine* in rat heart via α_1-adrenoceptors; however, the inhibition by unlabelled noradrenaline of the release of ^3H-*noradrenaline* was exclusively mediated by α_2-adrenoceptors in the same study. Finally, the inhibitory effect of methoxamine in the rat kidney was abolished by indomethacin, indicating an indirect, prostaglandin-mediated mechanism (Rump and Majewski 1987; see, however, rat heart: Hicks et al. 1986).

Further overflow studies are needed. If presynaptic α_1-receptors are confirmed, are they in fact targets of released noradrenaline and, hence, autoreceptors *sensu stricto?* The possibility of indirect, e.g. prostaglandin- or adenosine-mediated, drug effects deserves special attention (see also Hedqvist 1981).

3.3 Phenylethylamine and Imidazoline Recognition Sites?

We normally assume that all agonists bind to (almost) the same area of a given receptor. One consequence is that despite varying affinities and efficacies that result from each compound's specific mode of attachment and detachment, they are all equally sensitive to blockade by competitive antagonists. However, it has been proposed that α-adrenoceptors, including presynaptic α_2-autoreceptors, possess two recognition sites, one for phenylethylamine agonists and one for imidazoline agonists, with sufficient differences to entail functional consequences. This would be a first nonhomogeneity *within* the group of presynaptic α_2-autoreceptors. A note on the hypothesis in general will precede its discussion in respect of α_2-autoreceptors.

The two recognition site hypothesis in general, applicable to both α_1- and α_2-adrenoceptors, was developed by Ruffolo and his colleagues (see Ruffolo 1984; and Savola et al. 1986 for a recent study). It is based mainly on structure-activity relationships but also on such distinguishing features as selective desensitization without cross-desensitization, the use of different calcium sources by the two groups of agonists, and different susceptibility to competitive antagonists.

However, these distinguishing features are not found consistently, and it is by no means clear whether they in fact differentiate phenylethylamines (such as catecholamines) from imidazolines (clonidine, oxymetazoline; some nonimidazolines such as xylazine and B-HT 920, it has been suggested, also belong to this group). Selective desensitization and calcium dependence may perhaps rather separate agonists of low efficacy from those of high efficacy (Kenakin 1984). At the α_1- and α_2-adrenoceptors of smooth muscle cells, competitive antagonists, more often than not, act with the same potency against phenylethylamines and imidazolines (e.g. Sakakibara et al. 1982; Digges and Summers 1983; Kenakin 1984; Alabaster et al. 1985; Honda et al. 1985; Polónia et al. 1985; Tayo et al. 1986). Competitive antagonists also act with similar potency against noradrenaline and clonidine at the presynaptic α_2-receptors of serotonergic axons (Schlicker et al. 1983) and, probably, at the inhibitory α_2-adrenoceptors of the cholinergic neurons of the guinea-pig ileum (compare, for instance, Drew 1978b with Grundström et al. 1981). Anomalies in the interaction of α-antagonists with catecholamines have sometimes been observed in the ileum (Malta et al. 1981; Mottram 1983; Bond et al. 1986), but these may have alternative explanations (Malta et al. 1981). Phenylethylamine and imidazoline agonists may bind to distinct receptor areas; however, the functional consequences of this are, on the whole, not clear.

The view that presynaptic α_2-autoreceptors also possess two binding areas originated from a closer look at an early finding. It had been observed that cocaine reduced or even abolished the inhibition by oxymetazoline and clonidine of the electrically evoked release of noradrenaline in isolated tissues (e.g. Starke 1972a; Starke and Altmann 1973). The explanation had been that cocaine blocked the reuptake and increased the biophase concentration of noradrenaline, that autoinhibition then became near-maximal, and that under these conditions an exogenous agonist had little opportunity for further inhibition. These results were confirmed by Pelayo et al. (1980) and Langer and Dubocovich (1981). However, reinterpretation was suggested by additional observations, of which an important one was that blockade of reuptake did *not* reduce the release-inhibiting effect of α-methylnoradrenaline. What, if not an increase in perineuronal noradrenaline (which should have reduced the effect of α-methylnoradrenaline as well), was responsible for the loss of the imidazoline effect after uptake blockade? It is at this point that the idea of distinct binding sites was brought into play, and on its basis Langer and colleagues suggested two possibilities. First, uptake blockers might interact directly with the imidazoline recognition site of presynaptic α_2-autoreceptors and thus selectively decrease the effect of imidazolines (Langer and Dubocovich 1981). Second, the uptake blockers might act primarily at the uptake mechanism, and this interaction might then "somehow" disturb the effect of imidazolines but not phenylethylamines (Pelayo et al. 1980). This latter possibility has gained further support (Göthert et al. 1983).

As is often the case, the "truth" may be a composite of the initial explanation (see also Sect. 6.1) and the provocative idea of a more direct uptake site-presynaptic receptor interaction. How this interaction might be envisaged is shown in Fig. 2. Cocaine, desipramine, amphetamine, maprotiline, and a decrease in temperature all depress the uptake of noradrenaline. All of them, including low temperature (Göthert et al. 1983), "somehow" prevent imidazoline agonists but not phenylethylamines (which act at a different but partly overlapping area of the receptor) from triggering the release-inhibiting mechanism. The mystery of the "somehow" is indicated by the question mark. Figure 2 incorporates a proposal for the binding of α_2-adrenoceptor antagonists. Their attachment site is positioned such as to permit antagonism against both phenylethylamines and imidazolines. As mentioned above, these two agonist groups are similarly susceptible to competitive blockade at the α_2-receptors of smooth muscle, serotonin axons, and cholinergic neurons. Table 3 indicates that they are also similarly susceptible to competitive blockade at α_2-autoreceptors. In some postsynaptic response studies on rat vas deferens, α-receptor-blocking agents antagonized competitively imidazolines but not catecholamines (Mottram 1983; Vizi et al. 1983; Hicks et al. 1985); other authors, however, obtained competitive antagonism against noradrenaline as well (Tayo 1979; Leedham and Pennefather 1982; Warming et al. 1982). Partly overlapping sites for phenylethylamines, imidazolines, and antagonists are thus compatible with much of the available evidence.

How realistic, then, is an arrangement of the kind in Fig. 2? To appreciate the implications of the question, we should for a moment look beyond the α_2-autoreceptors. Uptake blockade interferes with other presynaptic receptors as well. The serotonergic axons innervating the rat hypothalamus possess inhibitory serotonin autoreceptors.

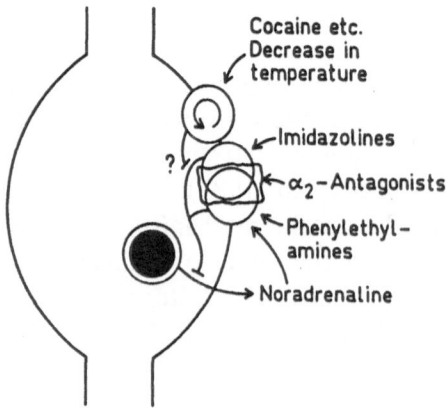

Fig. 2. A noradrenergic varicosity and the possible (controversial) interaction between noradrenaline transport sites and presynaptic α_2-autoreceptors

Serotonin, LSD, and 5-OCH$_3$-T all act there to depress the release of serotonin. In an impressive analogy with α$_2$-autoreceptors, serotonin uptake blockers interfere with the effect of LSD and 5-OCH$_3$-T (Galzin et al. 1985) but not of serotonin itself (as shown by the fact that the uptake blockers *enhance* the autoinhibition by endogenous serotonin). Can the analogy be pursued to a two-recognition-site model with an uptake blockade-sensitive area for LSD and 5-OCH$_3$-T, and an uptake blockade-insensitive area for serotonin at the serotonin autoreceptor? The noradrenaline axons of the rabbit hypothalamus possess inhibitory dopamine receptors. Only the effect of apomorphine and pergolide but not that of dopamine itself disappears when the uptake of noradrenaline is blocked (Galzin et al. 1982; Galzin and Langer 1985; see Göthert et al. 1983 for a fourth example). Are there different binding areas for apomorphine and pergolide on the one hand and dopamine on the other? Are transmitter uptake sites and presynaptic receptors "somehow" coupled in several species of neurons?

It seems fair to say that the answer is not yet known. An alternative should be mentioned. As pointed out above, some general differences between the effects of phenylethylamines and imidazolines, first ascribed to different recognition sites, were later explained by different agonist *efficacies*. In the case of presynaptic α$_2$-autoreceptors, Göthert et al. (1983) suggested that it might be the lower *efficacy* of clonidine as compared with noradrenaline (Medgett et al. 1978) that makes the former selectively sensitive to uptake blockade. One might speculate, for instance, that uptake blockade inactivates only a fraction of α$_2$-autoreceptors so that sufficient receptors remain intact for an effect of high-efficacy agonists (phenylethylamines), but not for an effect of low-efficacy agonists (imidazolines). The speculation might easily be extrapolated to other presynaptic receptors. In the end, the "different recognition site" view and the "different efficacy" view may amount to the same, because the efficacy of an agonist *results from* its specific mode of binding to the receptor. Even so, efficacy would be the decisive functional property that distinguishes uptake blockade-sensitive from uptake blockade-resistant presynaptic agonists. The possibility of measuring presynaptic efficacies and receptor reserves (Fuder et al. 1986) makes such speculation open to verification or disproof.

Finally, dissenting opinions and an encouraging perspective will be considered. All studies on the uptake mechanism-presynaptic receptor interaction mentioned thus far were carried out on isolated but otherwise intact tissues, and electrical pulses were used to trigger transmitter release. However, when Raiteri et al. (1984) examined the interaction on synaptosomes from rat brain and used high K$^+$ as a stimulus, they

found that neither was the effect of clonidine on the release of noradrenaline changed by cocaine and desipramine, nor was the effect
of LSD on the release of serotonin changed by serotonin uptake inhibitors. Is the arrangement of carrier sites and receptors (Fig. 2)
destroyed in synaptosomes? Or is only the modulation of electrically
evoked release — and not the modulation of K^+-evoked release — sensitive
to uptake blockade (Galzin et al. 1986)? Against the latter idea stands
the repeated observation that cocaine also does not impede the presynaptic inhibitory effect of imidazolines in the *electrically* stimulated
rat vas deferens (Moore and Griffiths 1982; Warming et al. 1982; Fiszman
and Stefano 1984). Or, finally, is Fig. 2 wrong and is the biophase
concentration of released transmitter (which is negligible in superfused
synaptosomes; see Sect. 3.1), after all, the only link through which
uptake blockers interfere with the effect of presynaptic autoreceptor
agonists (see also Bonanno and Raiteri 1987)?

A clonidine-displacing substance (CDS) has recently been partially
purified from calf brain. It is distinct from noradrenaline, displaces
^{3}H-clonidine and ^{3}H-yohimbine from their specific membrane binding
sites, and produces yohimbine-sensitive inhibition of electrically evoked
contractions of rat vas deferens (Atlas and Burstein 1984; Diamant
and Atlas 1986). Like clonidine, it lowers arterial pressure when injected
into the rostral ventrolateral medulla in rats (Meeley et al. 1986), but
surprisingly and in contrast to clonidine, it increases blood pressure
when injected into the same brain region in cats (Bousquet et al. 1986).
Thus, we can conclude this section with a final question: do phenylethylamine and imidazoline recognition sites, if indeed they exist as
functionally different sites, have their own endogenous ligands, namely,
noradrenaline and CDS, respectively?

3.4 Other Kinds of Nonhomogeneity?

There may be species and tissue differences in presynaptic α_2-auto-
receptors (e.g., Lattimer and Rhodes 1985), but they have not yet led
to a general further subdivision. α_2-Adrenoceptors, like other receptors,
may occur in a low- and a high-affinity state, perhaps not only for
agonists but also for antagonists (Asakura et al. 1985).

Not all presynaptic α_2-autoreceptors may be exposed to the same
concentration of released noradrenaline. Some will be close to, while
others will be more remote from, the sites of release. Hence, more
receptors may be available for exogenous agonists than for endogenous
released noradrenaline (see also Angus et al. 1984). Such an arrangement,
which is not unlikely in view of what has been learned about adreno-

ceptor distribution, might explain why in some studies the partial agonists clonidine (Medgett et al. 1978) and phentolamine (Ennis and Lattimer 1984; Limberger and Starke 1984) inhibited the release of noradrenaline at low concentrations but did not change it, or enhanced it, at high concentrations (Duckles 1982; Stevens and Moulds 1982; Ennis and Lattimer 1984; Kalsner 1985a). Possibly, at low degrees of receptor occupation the agonist effect prevails (for which *all* receptors are available), whereas at high degrees of receptor occupation the antagonist effect predominates (which occurs only at the receptors within reach of endogenous noradrenaline).

In some recent overflow studies, yohimbine seemed to be more potent in enhancing the release of noradrenaline than in antagonizing the inhibitory effect of exogenous α-agonists (Kalsner 1982, 1985a; Baker et al. 1984). Difficulties in attempts to block presynaptic inhibitory effects of α-agonists have also been encountered in electropharmacologic experiments (e.g., Suzuki 1984; see, however, Blakely et al. 1981; Illes and Starke 1983; Nörenberg 1986). The nonhomogeneities discussed here are a possible reason.

There may be both α_1- and α_2-receptors at some noradrenergic axons. α_2-Autoreceptors themselves may differ between species and tissues, interact differently with major agonist classes, change their affinity state, and be located at varying distances from the sites of release. All this should not distract us from the fact that the main presynaptic α-autoreceptor is the α_2-receptor, a tolerably well-defined entity as befits the prototype of a new pharmacologic receptor group, and the common site where exogenous α_2-adrenoceptor agonists and antagonists modulate the release of noradrenaline.

4 Receptor Location

The presynaptic α_2-autoreceptor has been represented in Fig. 2 as part of the varicosity membrane. This position is plausible but speculative. We do not even know for sure whether the receptor is a constituent of the noradrenergic axon at all, let alone where in the axon it is located. Like previous definitions (Starke and Langer 1979), the definition of presynaptic α_2-autoreceptors given above has therefore been deliberately noncommittal with respect to topography. Seven considerations and findings, however, support a direct presynaptic location.

1. α-Adrenoceptor agonists and antagonists modulate the release of noradrenaline in a variety of tissues with a variety of perineuronal

environments. The most parsimonious hypothesis is receptor location on the one common element, namely, the noradrenergic axons themselves.

2. Noradrenaline neurons possess soma-dendritic α_2-autoreceptors, almost certainly directly on the cell body and dendrites, thus indicating that these cells do express the gene coding for the α_2-receptor.

3. When the excretory ducts of the rat submaxillary gland are ligated, the secretory cells undergo atrophy. In contrast, the sympathetic axons and their α_2-adrenergic modulation survive. The modulation does not operate through the postsynaptic elements (Filinger et al. 1978).

4. In rat brain cortex slices, oxymetazoline reduces the release of noradrenaline elicited by high K^+ in the presence of tetrodotoxin. A tetrodotoxin-sensitive impulse traffic in neuronal loops (which might be preserved in the slices) is not necessary for the α-adrenergic modulation (Dismukes et al. 1977).

5. α-Adrenoceptor agonists depress the K^+-evoked release of noradrenaline from synaptosomes, and the agonist effect is counteracted by antagonists (de Langen et al. 1979; Ebstein et al. 1982; Raiteri et al. 1983, 1984). It is difficult to conceive an alternative to direct drug effects on the noradrenaline-releasing synaptosomes.

6. Lesions have been combined with radioligand binding experiments in the search for α_2 binding sites on noradrenergic axons. Often there was no change, or even an increase, in the binding of ^{3}H-clonidine or similar ligands after destruction of noradrenergic axons (e.g., Tanaka and Starke 1979; Groß et al. 1985; see U'Prichard 1984). Yet, a loss of binding sites has by now been reported at least five times (Story et al. 1979; Morris et al. 1981; Dooley et al. 1983; U'Prichard 1984; Tsukahara et al. 1986). All of these findings should be considered with caution because changes in postsynaptic α_2 binding sites must be expected after such lesions and make the interpretation equivocal (see Starke 1981b).

7. Once synthesized in the perikarya, α_2-autoreceptors should be transported down to the terminal axons. Laduron (1985) pointed out that such axonal transport of autoreceptors had never been demonstrated despite repeated attempts. However, he quoted only one published study, carried out on the hypogastric nerve of the cat (Alonso et al. 1982). In that study, the basic negative result was not failure to observe transport, but failure to detect any specific binding at all of ^{3}H-clonidine to membranes prepared from the nerve. No specific binding was found even in the spleen, a classic organ for α_2-adrenergic modulation. In other words, no specific binding was found where the receptor must exist,

wherever located, and this casts doubt on the suitability of the binding assay. An axonal transport of ^{3}H-clonidine binding sites has now been demonstrated in central noradrenergic neurons (Levin 1984).

5   The Mechanism Behind the Receptor

In his introduction to the session on transmitter release at the Second Meeting on Adrenergic Mechanisms, in Porto, Muscholl (1973) described the mechanism behind presynaptic inhibitory (muscarine) receptors as a "decrease (of) the availability of calcium ions for the stimulus secretion coupling process". This formula was subsequently often used and has probably guided many scientists' thoughts. *How* the "availability of calcium" is reduced, however, or in other words, what presynaptic receptors, in particular α_2-autoreceptors, actually do when activated has remained uncertain. Two main obstacles are that noradrenergic terminal axons are too small for a microelectrode and constitute too minute a fraction of the tissue for direct chemical analysis. Nevertheless, at least the questions asked and the hypotheses offered have become far more explicit in the last decade. The following discussion of questions and hypotheses will start from structural aspects of release, proceed to ionic mechanisms, and close with the regulatory proteins possibly involved, and will thus approach the primary changes following α_2-autoreceptor activation step by step. Hypotheses about early events will always be compared with observations on later events. For a review of mechanisms behind presynaptic receptors in general see Illes (1986).

5.1   Structural Aspects

Figure 3 summarizes some structural aspects of the release of noradrenaline and translates them into possibilities of modulation, as first tried by Stjärne (1978). Fig. 3A shows the noninhibited release. The action potential propagates along five varicosities but elicits release only in three: release is intermittent. Direct evidence for this fundamental property of sympathetic transmitter release is recent. It was obtained when the electrical response of smooth muscle cells of the vas deferens to nerve stimulation was investigated in detail. Blakely and Cunnane (1979) detected transient peaks in the rate of depolarization (the rate at which the excitatory junction potential rises) which they called "discrete events" and which appeared to be the response to transmitter released from single nearby varicosities. Their second important observation was that discrete events with a fixed latency, representing release from one and the same varicosity, occurred not upon each stimulus but only every few stimuli, i.e., intermittently. There is some dispute about the degree of intermittency (release between once every 2 and once every 500 stimuli). There is also dispute about whether release is basically monoquantal or whether several quanta (1–10) are normally released by a "successful" stimulus (Blakely and Cunnane 1979; Blakely et al. 1984b; Cunnane and Stjärne 1984b). Moreover, the cause of the intermittency is unknown (see Cunnane and Stjärne 1984b; Ryan et al. 1985); it might be a block of action potential propagation at some critical site or, as in Fig. 3A, a failure despite noninterrupted conduction, perhaps because no vesicle is ready for secretion. Notwithstanding such uncertainties, however, three principal possibilities of modulation follow. First, activation of α_2-autoreceptors may induce a propagation block (shift a preexisting block to more proximal varicosities? Fig. 3B). Second, propagation may be unimpaired, but the degree of intermittency may be increased for another reason (Fig. 3C). Third, if release is multiquantal, the quantal content may fall (Fig. 3D). Which alternative holds true?

The role of an impulse conduction block (Fig. 3B) can be estimated when one compares the modulation of electrically evoked release (which depends on propagation) with the modulation of release evoked by high K^+ concentrations (which is tetrodotoxin-resistant and does not depend on impulse propagation). Exogenous α_2-agonists inhibit either kind of release to a similar degree, thus indicating that their major effect is *not* a conduction block. (In brain slices, part of the K^+-evoked release is prevented by tetrodotoxin. α_2-Agonists again inhibit the tetrodotoxin-resistant fraction as well; Dismukes et al. 1977; Wemer et al. 1981.)

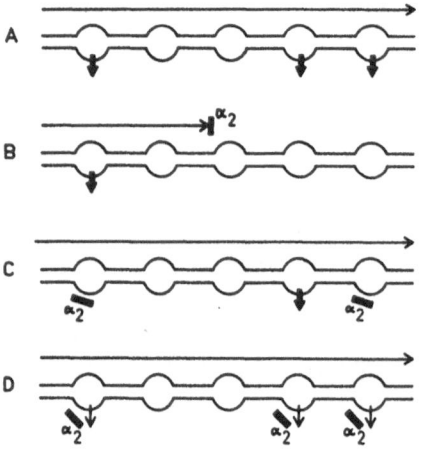

Fig. 3 A–D. Release of noradrenaline from a terminal varicose axon and possible modes of α_2-adrenergic inhibition. The action potential spreads from left to right, as shown by the horizontal *arrows*. A Noninhibited intermittent release. B α_2-Adrenoceptor activation blocks impulse propagation. The varicosity is a likely site of block, since the increase in fiber diameter reduces the safety factor for conduction. C α_2-Adrenoceptor activation increases the degree of intermittency (decreases the probability of release). D α_2-Adrenoceptor activation decreases the number of packets (quanta) released per active varicosity

When the comparison is extended to α_2-antagonists, however, differences show up. Stjärne and colleagues observed in guinea-pig vas deferens that the antagonists enhanced the electrically evoked release to a much greater degree than the K^+-evoked release; their conclusion was that *endogenous* noradrenaline causes *autoinhibition* mainly by blocking the spread of the action potential (Stjärne 1978; Alberts et al. 1981). The question is, however, whether this conclusion is generally valid (see Bevan et al. 1984; Ryan et al. 1985). In rat brain slices, phentolamine increased the propagation-independent part of K^+-evoked noradrenaline release at least as much as the propagation-dependent part (Fig. 6 of Wemer et al. 1981; compare also the effect of yohimbine on rat tail arteries in Galloway and Westfall 1982 and Westfall et al. 1985). Alberts et al. (1981) quote the failure of α-antagonists to increase K^+-evoked release from synaptosomes as support for a predominant propagation block by endogenous noradrenaline. However, this lack of antagonist effect is rather due to a very low perisynaptosomal concentration of released noradrenaline, as mentioned above (Sect. 3.1). A propagation block would not seem to be the generally prevailing mode of endogenous α_2-autoinhibition.

When Blakely and his coworkers studied the effects of clonidine, piperoxan, and yohimbine on discrete events, they found that the agonist increased the degree of intermittency *and* reduced the number of quanta released by a "successful" stimulus; the antagonists exerted the opposite effects (Blakeley et al. 1982, 1984b, 1986). The most likely conclusion to be drawn from the sum of all these findings is that there is not one exclusive manner in which α_2-autoreceptors modify the function of noradrenergic terminal axons. The receptors rather seem to have at their disposal all of the three modes of operation illustrated in Fig. 3B–D and to prefer perhaps different modes in different tissues. Hypotheses about ionic and enzymatic mechanisms should be examined for compatibility with each of the three structure-based modes of operation.

5.2 Ions

Presynaptic α_2-autoreceptors do not primarily affect the movement of Na^+ or Cl^- ions (Alberts et al. 1981; Wemer et al. 1981; Stjärne et al. 1986a). Changes in K^+ or Ca^{2+} are the ionic basis for the inhibition of noradrenaline release. Three primary changes have been proposed, namely, an opening of K^+ channels and an ensuing outward K^+ current and hyperpolarization; an inhibition of voltage-sensitive Ca^{2+} channels and an ensuing decrease of the depolarization-induced entry of Ca^{2+} ions

(I_{Ca}); and a membrane current-independent interference with some step between Ca^{2+} entry and exocytosis.

The first two hypotheses will be discussed together. It seems appropriate to begin with electropharmacologic studies on nerve cell bodies in which each of the two α-adrenergic membrane conductance changes has been demonstrated beyond doubt. α-Adrenoceptor activation produces a primary increase in K^+ conductance in locus ceruleus neurons of the rat (Andrade and Aghajanian 1985; Williams et al. 1985), neurons of the guinea-pig myenteric and submucous plexus (Morita and North 1981; North and Surprenant 1985), and perhaps rat, rabbit and frog sympathetic ganglia (Brown and Caulfield 1979; Cole and Shinnick-Gallagher 1981; Rafuse and Smith 1986). The receptors have been unanimously identified as $α_2$, most convincingly perhaps in vitro in the locus ceruleus (Williams et al. 1985) and the submucous plexus (North and Surprenant 1985). A primary α-adrenergic depression of I_{Ca} has been shown no less clearly, namely, in sympathetic ganglion cells (Horn and McAfee 1980) and locus ceruleus neurons of the rat (Williams and North 1985), as well as in dorsal root ganglion cells from chicken embryos (Dunlap and Fischbach 1981). The pharmacology of the receptors involved is not uniform, however. The receptor was classified as $α_2$ in rat sympathetic ganglia (McAfee et al. 1981) and as $α_2$-like, with some unusual properties, in chicken sensory neurons (Canfield and Dunlap 1984), but differed markedly from the $α_2$-type in the locus ceruleus (Williams and North 1985).

At nerve cell bodies, then, including the cell bodies of noradrenergic neurons, α-adrenoceptors can be coupled to both K^+ and Ca^{2+} channels. Can this be extrapolated to the nerve terminals? And which of the two kinds of channel modulation, if either, follows the activation of presynaptic $α_2$-autoreceptors – a primary opening of K^+ channels (Andrade and Aghajanian 1985; Williams et al. 1985) or a primary inhibition of voltage-dependent Ca^{2+} channels (Horn and McAfee 1980; Canfield and Dunlap 1984)? Either of the two might produce the modes of operation illustrated in Fig. 3B–D. A primary opening of K^+ channels and the ensuing hyperpolarization might impair impulse conduction (Fig. 3B; Stjärne 1978). However, it might also counteract the depolarization of single varicosities only and thereby, *secondarily,* reduce the entry of Ca^{2+} and prevent, or decrease the quantal content of, release (Fig. 3C, D); at the distal side of the varicosity, the decrease in fiber diameter might facilitate impulse conduction and allow further spread of the action potential (Ryan et al. 1985). A primary inhibition of Ca^{2+} channels would reduce the probability and quantal content of release (Fig. 3C, D). Yet it might also block impulse propagation (Fig. 3B) if the action potential were in part carried by I_{Ca} in the terminal axons, as may be the case (Cunnane and Stjärne 1984a). Which of the two competing hypotheses, if either, holds true? The following eight observations seem to be relevant. They have led many authors who worked on terminals rather than nerve cell bodies to favor the hypothesis of a primary decrease in I_{Ca}.

1. Local infusion of clonidine reduces the electrical excitability of terminal noradrenergic axons of the rat brain (Nakamura et al. 1981). Although hyperpolarization would be a likely mechanism (Nakamura et al. 1981), consistent with an increase in K^+ efflux, a suppression of I_{Ca} might be an alternative if a Ca^{2+} current contributed to the action potential.

2. The α_2-adrenergic hyperpolarization of nerve cell bodies is reduced after a decrease of the transmembrane K^+ gradient, as expected from an increase in K^+ conductance (Brown and Caulfield 1979; Morita and North 1981; Andrade and Aghajanian 1985). In contrast, the α_2-adrenergic presynaptic inhibition in the vas deferens was enhanced after a decrease of the K^+ gradient, thus indicating a different ionic mechanism (Illes and Dörge 1985).

3. One of the principal observations on presynaptic autoreceptors has been that when activated, they depress not only the electrically evoked release of noradrenaline but also the release that follows the depolarization produced by a rise in extracellular K^+. The inhibition of K^+-evoked release can easily be understood from the I_{Ca} hypothesis. Can it also be explained by a primary opening of K^+ channels? Theoretical calculations have given an affirmative answer (Dismukes et al. 1977; North and Williams 1984; see also the combined electrophysiological and biochemical experiments on opioid inhibition of K^+-induced noradrenaline release carried out by Bug et al. 1986). However, there is a discrepancy between theoretical predictions and experimental findings. The calculations predict that the inhibitory effect of hyperpolarization should disappear when the K^+ concentration used to trigger release is very high. At first glance, this seems to be the case (Starke and Montel 1974; Dismukes et al. 1977). In brain slices, for instance, oxymetazoline reduced the release of noradrenaline elicited by K^+ (26 mmol/liter) but did not change the release elicited by K^+ (56 mmol/liter). However, this pattern was found only as long as the concentration of Ca^{2+} was kept constant at 1.2 mmol/liter. When the concentration of Ca^{2+} was lowered to 0.2 mmol/liter, oxymetazoline did depress the release elicited by K^+ 56 mmol/liter (Dismukes et al. 1977). The operation of α_2-adrenergic inhibition at very high K^+ concentrations, provided only that the concentration of Ca^{2+} is sufficiently low, stands in marked contrast to predictions from the K^+ conductance hypothesis. Can the I_{Ca} hypothesis account for it?

The question can be generalized, and this underscores its importance. It has been found repeatedly that the modulation of noradrenaline release by presynaptic α_2-adrenoceptors (and by other receptors as well) is small or disappears when "strong" stimuli are used for release, e.g., very high K^+ as just discussed, high frequencies of electrical stimulation, pulses of long duration, and addition of drugs that promote release. It has also been found repeatedly that these strong stimuli become sensitive to modulation again when the extracellular Ca^{2+} concentration is reduced. Once more, can the I_{Ca} hypothesis of presynaptic modulation account for this general pattern? The answer probably follows from a consideration of axoplasmic Ca^{2+} as a function of both stimulus "strength" and the extracellular Ca^{2+} concentration. A common property of strong stimuli is that they raise the axoplasmic Ca^{2+} to high levels at which presumably the "release receptors" for Ca^{2+} are saturated (Stjärne 1973). Under these conditions, any modulation of release that is mediated by a modulation of the axoplasmic Ca^{2+} level, as for instance by a direct modulation of I_{Ca}, will necessarily be minimized; the excess of axoplasmic Ca^{2+} will override Ca^{2+}-modulating kinds of presynaptic modulation of release (Starke 1977; Westfall 1977; Vizi 1979; high axoplasmic Ca^{2+} may in addition desensitize the release receptors for Ca^{2+} or inactivate Ca^{2+} channels, and this may contribute to the loss of modulation). As soon as the external Ca^{2+} concentration is sufficiently lowered, however, the axoplasmic level can fall below the saturating range, and susceptibility to modulation is regained. The interdependence of stimulus "strength" and Ca^{2+} concentration does not by itself prove, but is at least compatible with, a primary modulation of I_{Ca}.

This is the place for a general caveat about the interpretation of autoreceptor studies. One often observes that an experimental manipulation (such as a change in the concentration of K^+ used for release) blunts or potentiates autoreceptor agonist or antagonist effects, and one may wish to draw conclusions from this change. If the manipulation itself has altered the release of noradrenaline, however, the interpretation is always ambiguous. The change in agonist or antagonist effect may have been due, first, to the change in biophase concentration of released noradrenaline resulting from the altered release; second, the general stimulus strength-dependence of modulation that reflects the axoplasmic Ca^{2+} level; or third, another reason such as a change in cyclic nucleotides (see below) — and it may be this third reason upon which one's interest is focused. No conclusion is possible without differentiation. Dismukes et al. (1977), in their study quoted above, did differentiate. They showed that a high biophase concentration of noradrenaline was *not* the factor that rendered release by K^+ (56 mmol/liter) resistant to oxymetazoline.

They also ruled out some other possible reasons and, most importantly, demonstrated that simply a decrease in Ca^{2+} restored the modulation. It is, in fact, a primary inhibition of voltage-dependent Ca^{2+} channels that explains best the α_2-adrenergic inhibition of K^+-evoked noradrenaline release.

4. TEA and 4-AP greatly enhance the release of noradrenaline and at the same time reduce or abolish the presynaptic inhibitory effect of α_2-agonists (Stjärne 1978; Wemer et al. 1981; Schoffelmeer and Mulder 1983a; Wakade and Wakade 1983). The two compounds block K^+ channels, and it might be supposed that they prevent a primary opening of K^+ channels by the α_2-agonists, in accordance with the K^+ conductance hypothesis. However, TEA and 4-AP *secondarily*, by prolonging the action potential, increase markedly the entry of Ca^{2+} into the terminal axons. This is how they increase the release of noradrenaline, and α_2-agonists might lose their effect also because a primary inhibition of I_{Ca} is rendered negligible when the axon terminals are flooded with Ca^{2+}. Low Ca^{2+} experiments argue against the former and support the latter possibility. In keeping with the pattern discussed in the preceding paragraphs, the inhibition caused by oxymetazoline, noradrenaline and adrenaline was restored when the external Ca^{2+} concentration was reduced (Wemer et al. 1981; Schoffelmeer and Mulder 1983a; Wakade and Wakade 1983). The α_2-agonists differed in this respect from acetylcholine and adenosine which remained ineffective in the presence of TEA even at low Ca^{2+}. Wakade and Wakade (1983) concluded that the α_2-agonists inhibit the "availability of calcium" directly, possibly by a decrease in I_{Ca}, whereas acetylcholine and adenosine primarily affect an electrical event such as the duration of the action potential.

5. The inhibitory effect of α_2-agonists — and often the facilitatory effect of antagonists as well (see Sect. 6.1) — declines with increasing frequency of stimulation. One reason for this is an increasing biophase concentration of noradrenaline (in the case of the agonists). A second reason, however, is the rise of intraaxonal Ca^{2+} to saturation levels. In support of this view, it has been observed that agonists and antagonists regain their effect when external Ca^{2+} is lowered (Langer et al. 1975).

6. The degree of α_2-adrenergic inhibition varies inversely with the external Ca^{2+} concentration (overflow measurements: de Langen and Mulder 1980; Alberts et al. 1981; Ebstein et al. 1982; Saelens and Williams 1983; Galzin et al. 1985; see, however, Kalsner 1981; postsynaptic response measurements: Drew 1978a; Magnan et al. 1979; Illes and Dörge 1985; see, however, Blakeley et al. 1984b, 1986). The relationship has generally been interpreted in terms of a primary in-

hibition of I_{Ca} (see the theoretical treatment by Kato et al. 1985) and is difficult to explain by a primary increase in K^+ conductance.

7. The dihydropyridine derivative Bay K 8644 promotes the entry of Ca^{2+} through voltage-dependent channels. Both Bay K 8644 and phentolamine increased the release of noradrenaline in rat vas deferens. The effects were not additive. The authors concluded that both drugs acted by an increase in I_{Ca} — Bay K 8644 by its direct action on the channel, and phentolamine by antagonism against an I_{Ca}-inhibiting effect of endogenous noradrenaline (Ceña et al. 1985).

8. A final group of experiments of some discriminating power was carried out on rabbit hearts that were perfused, and rat brain cortex slices that were superfused, with Ca^{2+}-free medium containing a depolarizing concentration of K^+. A brief introduction of Ca^{2+} elicited a release of noradrenaline, presumably because Ca^{2+} passed through the open channels, and in either tissue this "Ca^{2+}-induced release" was inhibited by α-agonists and enhanced by phentolamine (Göthert 1977; Göthert et al. 1979). The results in rabbit hearts speak perhaps most persuasively for a primary effect on Ca^{2+} and against a primary effect on K^+ channels, because the perfusion fluid contained no less than 140 mmol/liter K^+, making impossible any K^+ outward current and, hence, any modulation by a change in K^+ outward current (Göthert 1977).

Only those soma-dendritic α-receptors that open K^+ channels — and not those that inhibit Ca^{2+} channels — consistently display the properties of $α_2$-adrenoceptors, as mentioned above. It has been proposed that a given receptor subtype is always associated with the same ion channel modulation (North 1986). This unifying idea would make the opening of K^+ pores the most attractive ionic basis of presynaptic $α_2$-adrenergic inhibition. Yet direct observations on release from terminals favor a primary effect on Ca^{2+} channels. Perhaps the same receptor type can trigger different signal transduction chains at neuronal cell bodies and axon terminals.

The last of the three competing ionic hypotheses of $α_2$-presynaptic inhibition remains to be considered. Two observations on rat brain slices led Mulder, Schoffelmeer, and their colleagues to suggest recently that $α_2$-adrenoceptor agonists, rather than inhibiting transmembrane Ca^{2+} influx, might "inhibit some intracellular process distal to Ca^{2+} entry ..., for example, ... the binding of Ca^{2+} to some intracellular site(s) essential to the secretory mechanism" (Schoffelmeer and Mulder 1983a); or at least, that inhibition of Ca^{2+} entry and inhibition of the intra-axonal utilization of Ca^{2+} might operate simultaneously (Mulder et al. 1984; Mulder and Schoffelmeer 1985). The first observation was that there was a difference between the effects of α_2-ligands on the one hand and of the Ca^{2+} channel blocker cadmium on the other hand, on the release of noradrenaline (Schoffelmeer and Mulder 1983a). The second observation originated from the finding that veratrine released noradrenaline even in Ca^{2+}-free media containing EGTA; veratrine apparently elicited an influx of Na^+, which then mobilized Ca^{2+} from intracellular stores (Schoffelmeer and Mulder 1983b). Schoffelmeer and Mulder (1983c) subsequently reported that this veratrine-induced release *in the absence of extracellular Ca*$^{2+}$ was reduced by clonidine and enhanced by phentolamine, in a situation where any modulation of an inward Ca^{2+} current was ruled out.

The findings, particularly the second, more direct one, and the conclusion are intriguing. The conclusion is as compatible with many of the eight observations outlined above as is a primary inhibition of I_{Ca}. Questions, however, remain. Would a primary change occurring *after* the entry of Ca^{2+} account for the three alternatives in Fig. 3B—D? While there is no difficulty in the case of Fig. 3C and D, it is not easy to see how the change could block the propagation of the action potential (Fig. 3B). How would a primary change *after* the entry of Ca^{2+} reduce the electrical excitability of the terminal axons (Nakamura et al. 1981)? A third question concerns the release of noradrenaline elicited by ouabain in the absence of Ca^{2+}, the mechanism of which resembles that of the veratrine-induced release in Ca^{2+}-free media (Schoffelmeer and Mulder 1983a). Is this ouabain-induced release modulated by α_2-ligands? The answer is not known for brain slices. In rat vas deferens, ouabain-induced noradrenaline release at zero Ca^{2+} was *not* changed by clonidine and phentolamine (Stute and Trendelenburg 1984). It should be noted that the effect of veratridine (and certainly of the alkaloid mixture veratrine) is complex, even in the absence of Ca^{2+} (see Palaty 1984; Stute and Trendelenburg 1984; Pizarro et al. 1986); several components would have to be taken into account in a complete study of the modulation of veratrine-induced release. A final consideration that raises some doubt is that an α_2-adrenergic block of Ca^{2+} utilization would make all the gratifying analogies with the cell bodies fortuitous, where Ca^{2+} and K^+ *currents* are the target for modulation.

There are arguments for each of the three ionic hypotheses, and each has its advocates. An investigation of the effect of α_2-ligands on synaptosomal radiocalcium uptake might be helpful but does not seem to have been reported (for dopamine autoreceptors see Gripenberg et al. 1980). Hypotheses on regulatory proteins must be examined for their compatibility with all three potential ionic mechanisms.

5.3   Regulatory Proteins

Cyclic nucleotides, guanine nucleotide-binding regulatory proteins (G proteins), and protein kinases have been implicated in the α_2-adrenergic modulation of noradrenaline release, as in many cellular transduction mechanisms. It has been suggested that in brain slices, α-agonists cause presynaptic inhibition by an *increase* in cyclic AMP formation (Dismukes and Mulder 1976; see also Ebstein et al. 1982). Another suggestion was that in the pineal gland, α-agonists act by an increase in cyclic GMP (Pelayo et al. 1978). For about 8 years it has been clear, however, that an early event following α_2-adrenoceptor activation in many cells is a *decrease* of cyclic AMP formation (see Jakobs et al. 1985). Other details of the transduction mechanism have also been identified. α_2-Adrenoceptors inhibit the adenylate cyclase via a G protein, abbreviated G_i or N_i, which consists of three polypeptides, α, β, and γ. Guanosine triphosphate and its analogs bind to the α-subunit. Cyclic AMP seems to produce most or all of its effects by activation of a cyclic AMP-dependent protein kinase which then phosphorylates specific proteins, including neuronal enzymes, neuronal ion channels or proteins that modulate them, and synaptic vesicle constituents (see Browning et al. 1985).

The idea that inhibition of adenylate cyclase might also be responsible for the α_2-adrenergic inhibition of noradrenaline release seems to have first been put forward explicitly by Wikberg (1979). Is there, then, an adenylate cyclase in noradrenergic axons, and does cyclic AMP enhance the release of noradrenaline? Do α_2-agonists work through inhibition of this presynaptic enzyme? Would cyclase inhibition be able to initiate the ionic changes and structure-based modes of operation discussed above?

Table 4 summarizes studies with three groups of compounds: cyclic AMP analogs, forskolin, which activates the catalyst adenylate cyclase directly, and the phosphodiesterase inhibitors IBMX, rolipram, AH 21–132 and ICI 63, 197. All increased the electrically evoked release of noradrenaline (a negative finding with forskolin: Fredholm and Lindgren 1986; for studies before 1978 see Starke 1977). 8-Bromo-cyclic GMP caused no change except in guinea-pig heart; moreover, the phosphodiesterase inhibitors in Table 4, except IBMX, block the hydrolysis of cyclic AMP more effectively than that of cyclic GMP, thus indicating an involvement of cyclic AMP rather than cyclic GMP. Further evidence is provided by the effect of β-adrenoceptor agonists. These compounds, which in general act by stimulation of adenylate cyclase, enhance the release of noradrenaline in most if not all tissues, and the enhancement is potentiated by phosphodiesterase inhibitors (Celuch et al. 1978;

Table 4. Effect of cyclic nucleotide analogs, forskolin, and phosphodiesterase inhibitors on electrically evoked release of noradrenaline

Tissue	Drug and release-enhancing concentration in μmol/liter	Comment [a]	References
	Cat		
Spleen	IBMX 500	1 Hz	Celuch et al. 1978
	Rat		
Vas deferens	Forskolin 0.24	1 Hz	Hovevei-Sion et al. 1983
Cerebral cortex	Dibutyryl-cAMP 100 − 1000 8-Bromo-cAMP 30 − 1000 Forskolin 0.1 − 30 IBMX 1 − 1000 Rolipram 10 − 1000 AH 21−132 10 −100	1 − 2 Hz. No effect of 8-bromo-cGMP 100 − 1000	Schoffelmeer and Mulder 1983c; Markstein et al. 1984; Schoffelmeer et al. 1985a, b, 1986b
	Mouse		
Heart	ICI 63, 197 90 − 270	2 − 4 Hz	Johnston and Majewski 1986
	Guinea pig		
Heart	Dibutyryl-cAMP 0.001 8-Bromo-cGMP 0.01	Perfused heart. Endogenous noradrenaline and dopamine-β-hydroxylase. 5 Hz. Phenoxybenzamine	Langley and Weiner 1978
Ileum	Dibutyryl-cAMP 1000 − 4000 8-Bromo-cAMP 100 − 1000 Forskolin 10 − 100 IBMX 1000 − 4000	1 Hz. Desipramine, normetanephrine. No effect of 8-bromo-cGMP 500	Alberts et al. 1985
Vas deferens	8-Bromo-cAMP 100 − 1000 IBMX 500 − 2000	0.5 − 10 Hz. Desipramine, normetanephrine. No effect of 8-bromo-cGMP 100 − 1000	Stjärne et al. 1979
	Rabbit		
Pulmonary artery	Dibutyryl-cAMP 100 −1000 8-Bromo-cAMP 100 − 1000 Forskolin 1 − 10 Rolipram 1 − 10 AH 21−132 10 − 100 ICI 63, 197 30	0.66 − 2 Hz. Cocaine, corticosterone, propranolol	Göthert and Hentrich 1984; Johnston and Majewski 1986

Table 4 (continued)

Tissue	Drug and release-enhancing concentration in μmol/liter	Comment [a]	References
Cerebral cortex	Forskolin 0.1 – 10 IBMX 1 – 1000 AH 21–132 10 – 100	3 Hz	Markstein et al. 1984
	Man		
Pulmonary artery	8-Bromo-cAMP 100 – 1000 Forskolin 0.1 – 10 AH 21–132 1 – 30	2 Hz. Cocaine, corticosterone	Hentrich et al. 1985

Only studies not summarized by Starke (1977) are entered, and only studies on isolated tissues in which release was determined as overflow.

[a] Unless stated otherwise, tissues were incubated or superfused, and the electrically evoked overflow was determined with the aid of ^{3}H-noradrenaline. Frequency and auxiliary drugs (not necessarily present in all experiments) are indicated.

Johnston and Majewski 1986) and forskolin (Hentrich et al. 1985), as is expected for a cyclic AMP-mediated effect. In contrast to electrically evoked release, the K^+-evoked release of noradrenaline was sometimes reduced by phosphodiesterase inhibitors (Ebstein et al. 1982; Wemer et al. 1982; Schoffelmeer and Mulder 1983c; see, however, Pelayo et al. 1978; Göthert and Hentrich 1986); the reason is not clear. On the whole, the results support, albeit indirectly, the existence of an adenylate cyclase in noradrenergic terminal axons. They also suggest that cyclic AMP is not a *sine qua non* for release but promotes the release elicited by the nerve action potential (see already Cubeddu et al.1975).

Do α_2-agonists act by inhibition of this cyclase, as indicated in Fig. 4? The evidence is again indirect and comes from two kinds of interaction experiment: first, the interaction of α_2-adrenoceptor ligands with cyclic AMP derivatives, forskolin, and phosphodiesterase inhibitors; and second, their interaction with compounds that uncouple the α_2-receptor from the cyclase.

Mulder and his colleagues argued that if presynaptic α_2-adrenoceptors were coupled to adenylate cyclase, any increase in axoplasmic cyclic AMP (or the introduction of cyclic AMP analogs) should reduce the degree of modulation. Accordingly, they studied the interaction of oxymetazoline, clonidine, and phentolamine on the one hand and of cyclic AMP analogs, forskolin, and rolipram on the other hand, in rat brain slices. Release of noradrenaline was elicited by high K^+ (Wemer et al. 1982), electrical pulses (Mulder and Schoffelmeer 1985), or veratrine at zero Ca^{2+} (Schoffelmeer and Mulder 1983c). Dibutyryl-cyclic AMP,

Fig. 4. A speculative molecular mechanism of α_2-adrenergic presynaptic inhibition. The varicosity possesses an adenylate cyclase (AC) that catalyzes the formation of cyclic AMP. Cyclic AMP in turn activates a protein kinase (PK); the pathway then branches. The kinase may phosphorylate a protein of, or close to, the voltage-dependent Ca^{2+} channel to increase the entry of Ca^{2+}. It may also phosphorylate synapsin I ($Sy\ I$) to enhance exocytosis. A Ca^{2+}/calmodulin-dependent protein kinase also phosphorylates synapsin I

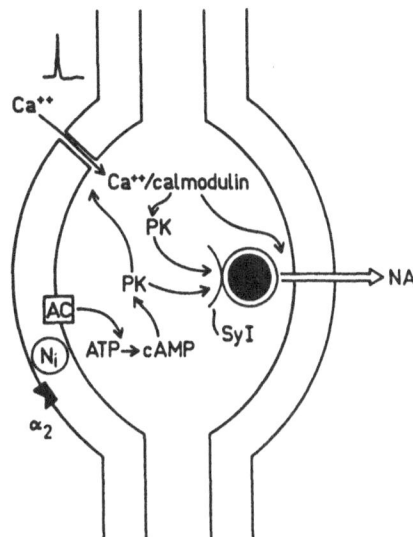

8-bromo-cyclic AMP, and forskolin consistently reduced the percentage inhibition caused by the agonists and the percentage increase caused by phentolamine. The interaction with rolipram was studied only on K^+-evoked release, which was depressed by the drug, as mentioned above; rolipram did not alter the effects of oxymetazoline and phentolamine (Wemer et al. 1982). The authors concluced that the pattern observed, except in the case of rolipram, supported cyclase inhibition as the mechanism behind presynaptic α_2-autoreceptors (see also another recent interaction experiment by Schoffelmeer et al. 1986b, in which release was elicited by a Ca^{2+} ionophore).

There are divergent results, however. In several studies on a number of tissues, including brain, forskolin and phosphodiesterase inhibitors (also where they increased release) did not change, or even augmented, the effects of α_2-adrenoceptor ligands (Cubeddu et al. 1975; Ebstein et al. 1982; Göthert and Hentrich 1984; Markstein et al. 1984; Alberts et al. 1985; Johnston and Majewski 1986). Cubeddu et al. (1975) and Alberts et al. (1985) expressly concluded that cyclic AMP is not a link in presynaptic α_2-adrenergic inhibition. It should also be remembered that any manipulation that increases the release of noradrenaline, any increase in the "strength" of the stimulus, irrespective of the mechanism involved, will curtail presynaptic modulation when the axoplasmic Ca^{2+} level becomes high (see above). An interaction at the level of the axoplasmic Ca^{2+} rather than cyclic AMP concentration does not seem to have been excluded.

The α_2-adrenoceptor can be uncoupled from the adenylate cyclase by compounds that interfere with the function of N_i, specifically its

α-subunit. Three groups of compounds have been employed in studies on presynaptic α_2-autoreceptors, namely, the sulfhydryl reagent N-ethylmaleimide, IAP, which is one of the toxins of *Bordetella pertussis* and catalyzes the NAD-dependent ADP-ribosylation of the α-polypeptide (see Bokoch et al. 1985; Jakobs et al. 1985), and phorbol esters; the phorbol esters that were used activate the Ca^{2+}/phospholipid-dependent protein kinase, also known as protein kinase C, which in turn phosphorylates the α-subunit of N_i (Katada et al. 1985). If α_2-autoreceptors rely on the N_i-cyclase translation machinery, these compounds should increase the release of noradrenaline (by interruption of autoinhibition) and blunt the modulation by exogenous α_2-ligands. Both predictions have been fulfilled (N-ethylmaleimide: Allgaier et al. 1986a; IAP: Lai et al. 1983; Allgaier et al. 1985; phorbol esters: Wakade et al. 1985, a report in which only an increase in noradrenaline release is shown; Allgaier and Hertting 1986a). For instance, the electrically evoked release of noradrenaline in rabbit hippocampal slices was increased by 23%–54% after 6 h of incubation with IAP. Simultaneously, the modulatory effects of clonidine and yohimbine were reduced (Allgaier et al. 1985).

Again, however, all of the uncoupling agents enhanced the release of noradrenaline themselves. This is the expected effect. Even so, it clouds the interpretation of the interaction with clonidine and yohimbine. In two recent studies, Allgaier and Hertting (1986b) and Allgaier et al. (1986b) in fact appear to suggest that phorbol esters do not affect the transduction system behind the α_2-autoreceptor. There is a second interesting element of ambiguity. IAP is now known to catalyze the ADP-ribosylation of other G proteins in addition to N_i, such as the brain protein G_O (Sternweis and Robishaw 1984). G proteins probably not only regulate the activity of adenylate cyclase but also take part in other signal transduction mechanisms (see Bokoch et al. 1985). The uncoupling experiments may indicate that a G protein is a link in presynaptic α_2-adrenergic inhibition. Yet whether this G protein is N_i and the enzymatic consequence inhibition of adenylate cyclase remains a matter of conjecture. A transduction mechanism involving a G protein but not the cyclase has recently been suggested for the inhibition of noradrenaline release through presynaptic adenosine receptors (Fredholm and Lindgren 1986) as well as for the inhibition of Ca^{2+} channels in chicken sensory ganglion cells through the α_2-like receptors mentioned above (Sect. 5.2; Forscher et al. 1986; Holz et al. 1986; Rane and Dunlap 1986). The two types of interaction experiment have not yet unequivocally verified or refuted the adenylate cyclase hypothesis of α_2 presynaptic inhibition.

If the presynaptic inhibition starts with an inhibition of adenylate cyclase, could this lead to the appropriate ionic changes and changes in the function of entire terminal axons? In principle, the answer is yes, but for any details we have to resort to speculation, some of which will conclude this chapter.

For instance, since cyclic AMP can inhibit neuronal K^+ channels (in invertebrates; see Browning et al. 1985), suppression of cyclic AMP formation might in fact promote the opening of K^+ channels. There is evidence to show that this may happen in noradrenergic neurons; the α_2-adrenergic hyperpolarization of locus ceruleus cells, which is due to an opening of K^+ channels, was abolished by IAP (Aghajanian and Wang 1986) and by cyclic AMP analogs (Andrade and Aghajanian 1985).

Secondly, since cyclic AMP can increase voltage-dependent Ca^{2+} currents (see Browning et al. 1985), an α_2-adrenergic fall in cyclic AMP might in fact decrease I_{Ca}, as outlined in Fig. 4.

Cyclic AMP or a fall in cyclic AMP might also modify the intra-axonal action of Ca^{2+}, this being the ionic mechanism proposed by Mulder, Schoffelmeer, and their colleagues (see Mulder and Schoffelmeer 1985). A speculative pathway has been incorporated into Fig. 4. One of the substrates of cyclic AMP-dependent protein kinase in nerve terminals is synapsin I, a neuron-specific protein that seems to control the function of transmitter storage vesicles. Interestingly, noradrenaline itself, acting via a presynaptic β-adrenoceptor, promotes the phosphorylation of synapsin I (Mobley and Greengard 1985). Synapsin I is also phosphorylated by two Ca^{2+}/calmodulin-dependent protein kinases. Although this has been shown for Ca^{2+}/calmodulin-dependent protein kinase II only, the phorphorylations might change synapsin I in such a way that exocytosis is enhanced (Browning et al. 1985). Ultimately, the Ca^{2+} kinase cascade and the cyclic AMP cascade might work hand in hand to amplify the effect of Ca^{2+}. An α_2-adrenergic cyclase inhibition would then remove the amplification and depress the intra-axonal action of Ca^{2+}.

How does activation of presynaptic α_2-autoreceptors reduce the "availability of calcium"? There is no definite answer. Perhaps it is even wrong to search for a single invariable mechanism. The reaction chains could differ in different tissues and activity states of the neurons. Perhaps they could differ even with the particular agonist. If phenylethylamine and imidazoline agonists initiated different reaction chains, this might explain the (postulated) selective sensitivity of the imidazolines but not the phenylethylamines to inhibition of neuronal uptake. Factors that have not been considered at all, such as an enhancement of the removal of Ca^{2+} from the axoplasm, may also be involved in α_2-adrenergic presynaptic inhibition.

6 Physiology

The question of whether presynaptic α_2-autoreceptors play a physiologic role can be aimed at obtaining both simple and complex answers. It can simply refer to whether, apart from exogenous drugs, endogenous noradrenaline also has access to the receptors, but can also refer to how great is the relative importance of autoinhibition, in comparison with other presynaptic receptors and changes in impulse flow, in determining synaptic transmission in vivo. Three aspects will be discussed. Does endogenous noradrenaline activate presynaptic α_2-autoreceptors under appropriate conditions in vitro? Do the changes in transmitter release lead to changes in the postsynaptic response? Can the function of peripheral α_2-auto-receptors in vivo be discovered by determination of noradrenaline overflow? As the discussion progresses, its focus will shift from the in vitro studies that form the basis of this essay to observations on whole animal preparations and, finally, intact animals and man.

6.1 Autoinhibition In Vitro

The basis for the idea of an endogenous α_2-autoinhibition is, of course, the almost uniform release-enhancing effect of α-adrenoceptor antagonists. The enhancement is mediated by a tolerably well-defined receptor, the prototype α_2-adrenoceptor, and it is hard to imagine that it could be due to anything but antagonism against an endogenous agonist. But we can use additional touchstones. The hypothesis predicts that the enhancement by antagonists (when given at sufficiently high concentrations) should *ceteris paribus* vary directly, and the inhibition by exogenous agonists vary inversely, with the biophase concentration of released noradrenaline. Can these predictions be verified? The *ceteris paribus* condition poses a problem. For instance, changes in stimulation frequency would be an obvious means of altering the biophase concentration of noradrenaline. Changes in frequency, however, like all changes in stimulus strength, also affect the postreceptor signal transduction machinery, presumably by way of the frequency-dependent axoplasmic Ca^{2+} level (see Sect. 5.2). The effect of an α_2-agonist may decline with increasing frequency, not because it has to compete with more released noradrenaline but because the axoplasmic Ca^{2+} level becomes too high. For the same reason the effect of an antagonist may decrease rather than increase with increasing frequency. The addition of cocaine-like drugs provides a second example. These compounds increase the perineuronal concentration of released noradenraline but may at the same time interfere with the effect of imidazoline agonists more directly (see Sect. 3.3). In either instance, the *ceteris paribus* condition is not met, and the results remain ambiguous. Findings will now be summarized in which these interpretational difficulties have been avoided and which support the interaction of agonists and antagonists in vitro with an ongoing, noradrenaline-mediated, and noradrenaline concentration-dependent autoinhibition.

First, five groups of observations on agonists and partial agonists.

1. The inhibitory effect of oxymetazoline varied inversely with the biophase concentration of noradrenaline, even when all experimental procedures were kept constant (Starke 1972a).

2. The effect of clonidine (in the presence of a low concentration of cocaine) decreased when the degradation of released noradrenaline was prevented by pargyline, an intervention that should leave the "strength" of the releasing stimulus unchanged (Reichenbacher et al. 1982).

3. Inhibition by imidazolines is reduced by noradrenaline uptake blockers, an observation of doubtful interpretation, as mentioned above. When, however, the concentration of cocaine was raised so high that it became local anesthetic and decreased the release of noradrenaline, clonidine regained its full effect, and a similar relationship was obtained for the phenylethylamine α-methylnoradrenaline (Reichenbacher et al. 1982). The biophase concentration of noradrenaline was the crucial determinant under these conditions.

4. Figure 5 shows a paradigm in which ³H-amezinium was used to estimate transmitter release from cerebrocortical noradrenergic axons. Amezinium is selectively taken up into noradrenaline storage granules and can subsequently be released. It does not possess affinity for adrenoceptors (see Hedler et al. 1983). Some rats were pretreated with α-MT in order to deplete their noradrenaline stores. In some brain slices from pretreated animals, the stores were refilled by incubation with unlabeled noradrenaline immediately after preparation. Other slices from α-MT rats were also incubated with unlabeled noradrenaline, but in the presence of cocaine in order to prevent refilling. The left-hand panel of Fig. 5 shows that the fractional release of ³H-amezinium in slices from α-MT-pretreated rats was more than twice that in slices from nonpretreated rats, thus indicating disinhibition due to depletion of noradrenaline. In support of this view, it was observed that release was low again when the noradrenaline stores had been refilled. Finally, prevention of refilling by cocaine prevented the re-establishment of autoinhibition. These findings alone already reveal a noradrenaline-mediated auto-

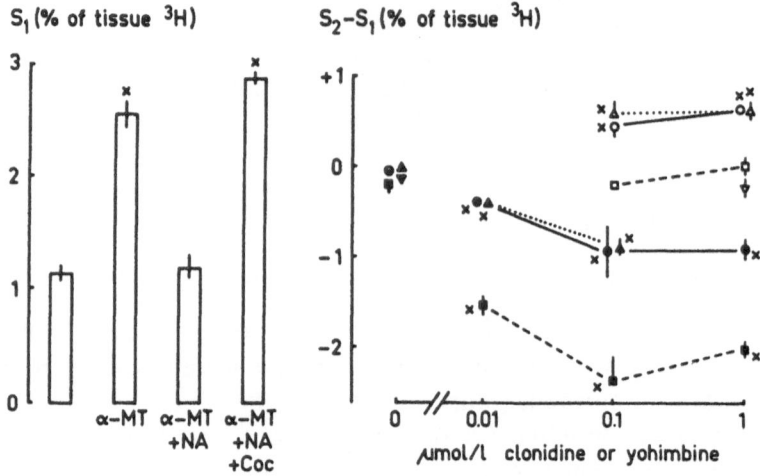

Fig. 5. Effect of clonidine and yohimbine on the evoked overflow of tritium from rat cerebrocortical slices preincubated with ^{3}H-amezinium. After preincubation, the slices were superfused with ^{3}H-amezinium-free medium. They were stimulated electrically twice for 3 min each at 3 Hz, after 60 and 95 min of superfusion (S_1, S_2). Clonidine or yohimbine was added 20 min before S_2. Pretreatments: some rats were pretreated with α-MT; some slices from α-MT rats were incubated with noradrenaline before the preincubation with ^{3}H-amezinium ($α\text{-}MT + NA$); some slices from α-MT rats were incubated with noradrenaline in the presence of cocaine before the preincubation with ^{3}H-amezinium ($α\text{-}MT + NA + Coc$). *Left-hand panel:* overflow of tritium evoked by S_1, expressed as a percentage of the tritium content of the tissue. *Right-hand panel:* concentration-response curves of clonidine (*filled symbols*) and yohimbine (*empty symbols*). The ordinate shows the difference in the overflows evoked by S_2 and S_1. ● and ○ show effects in slices from unpretreated rats; ■ and □ in α-MT slices; ▲ and △ in α-MT + NA slices; ▽ in α-MT + NA + Coc slices. Significant differences from unpretreated rats (*left-hand panel*) and corresponding controls (*right-hand panel*): × P < 0.005. Modified from Hedler et al. (1983)

inhibition. The right-hand panel of Fig. 5 demonstrates the effect of clonidine. It caused moderate inhibition in slices from unpretreated rats. After depletion, its effect was much more marked. Finally, replenishment of the stores brought the effect of clonidine back to normal, as expected for an interplay of exogenous inhibition and autoinhibition (see below for yohimbine).

5. The dependence on the biophase concentration of released noradrenaline becomes rather impressive when the exogenous agonist is a partial agonist. A presynaptic partial agonist should decrease release at low noradrenaline concentrations; it should increase release when noradrenaline is present at concentrations producing an inhibition greater than that induced by maximally effective concentrations of the partial agonist (Starke et al. 1974). This pattern was first achieved by Medgett et al. (1978) in guinea-pig atria. Clonidine reduced the release of

noradrenaline elicited by 5 pulses at 1 Hz (low biophase concentration of noradrenaline), did not change the release by 50 pulses at 2.5 Hz (intermediate noradrenaline concentration) but increased release by 150 pulses at 5 Hz (high noradrenaline). Phentolamine and idazoxan also behave as pre- and postsynaptic partial α_2-receptor agonists in some tissues. In the ear artery of the rabbit, a change from 13 pulses at 0.25 Hz to 52 pulses at 1 Hz converted idazoxan and phentolamine from inhibitory to facilitatory drugs (Limberger and Starke 1983, 1984; see also Medgett and Rand 1981; Duckles 1982; Heepe and Starke 1985). A thorough study with clonidine was recently carried out by Cichini et al. (1986). The temperature was varied as an additional procedure to alter the concentration of released noradrenaline. The metamorphoses that the effect of clonidine underwent led the authors to conclude that their "results strongly corroborate the concept of presynaptic regulation of noradrenaline release and emphasize the pivotal role of the biophase concentration of the released transmitter for the action of a partial agonist at the presynaptic receptor".

Second, six groups of observations on antagonists.

1. The release process in the nictitating membrane and spleen of the cat, when measured for instance as overflow of dopamine-β-hydroxylase, was increased after noradrenaline had been depleted by reserpine or α-MT. Simultaneously, the release-enhancing effects of phenoxybenzamine and phentolamine were reduced or abolished (Enero and Langer 1973; Cubeddu and Weiner 1975a; Dixon et al. 1979).

2. α-MT was also used in the ^{3}H-amezinium paradigm which will now be extended to yohimbine. The right-hand pattern in Fig. 5 shows that yohimbine increased the release of ^{3}H-amezinium in brain slices from nonpretreated rats. When the animals had been pretreated with α-MT, however, and release was disinhibited (left-hand panel), yohimbine failed to cause a further increase. Its effect returned with the refilling of the stores, and was absent once again when refilling was prevented by cocaine. Whenever noradrenaline was coreleased with ^{3}H-amezinium, release was low, only moderately reduced by clonidine, and disinhibited by yohimbine. Whenever very little noradrenaline was coreleased, release was high, clonidine had ample room for inhibition, and yohimbine remained without effect for lack of an agonist. The results extend those in cats mentioned in the preceding paragraph. They show that under appropriate conditions, as one manipulates the biophase concentration of released noradrenaline, one can manipulate the magnitude of release as well as the effect of exogenous agonists and antagonists, in accordance with the operation of an autoinhibitory system.

3. The rate of perfusion was changed in order to alter the biophase concentration of noradrenaline in the rat heart. The effect of yohimbine changed as predicted (Fuder et al. 1984).

4. Cocaine-like compounds may distinguish between imidazoline and phenylethylamine agonists but do not distinguish between imidazoline- and phenylethylamine-derived antagonists (Schlicker et al. 1983). Cocaine-like drugs have repeatedly been shown to augment the release-enhancing effect of α-adrenoceptor antagonists (Schlicker et al. 1983; Angus et al. 1984; Baker et al. 1984; Enero 1984; Fuder et al. 1984; Heepe and Starke 1985; Nedergaard 1986). In guinea-pig atria, for instance, phentolamine did not increase the release of noradrenaline elicited by 4–12 shocks at 0.25 Hz when given alone, but caused a marked increase when the biophase concentration of noradrenaline had been raised by desipramine (Angus et al. 1984).

5. Many experiments with stimulus trains of different length and frequency can also be adduced here, although it should be remembered that now the stimulus strength changes. This may confuse the picture, and discrepant results will have to be discussed below. Phentolamine had no effect in guinea-pig atria stimulated by 12 pulses at 0.25 Hz as just mentioned (in the absence of desipramine), but increased the release evoked by 12 pulses at 2 Hz when, owing to the shorter pulse intervals, the biophase noradrenaline concentration was much higher (Angus et al. 1984). In cerebrocortical slices from rabbits, idazoxan caused no change when 36 pulses were distributed over 120 s (0.3 Hz), caused a moderate increase when 120 pulses were applied over the same time (1 Hz), and caused a marked increase at 360 pulses within 120 s (3 Hz; Heepe and Starke 1985). Similar results were obtained with other antagonists and in other tissues (Cubeddu and Weiner 1975a; Baumann and Koella 1980; Story et al. 1981; Auch-Schwelk et al. 1983; Marshall 1983; Limberger and Starke 1984; Heepe and Starke 1985; McCulloch et al. 1985; Hieble et al. 1986; Nedergaard 1986). The pattern of increasing enhancement with increasing frequency is *the opposite* of the usual inverse frequency-modulation relationship, and the opposite also, most notably perhaps, of the decreasing enhancement of release with increasing frequency in the case of other presynaptic facilitatory drugs such as β-adrenoceptor agonists and adrenocorticotropic hormone (see Göthert 1984). The pattern singles out the α-adrenoceptor antagonists, in accordance with the view that it is only these drugs that produce enhancement as antagonits to released noradrenaline.

6. A number of postsynaptic response studies have yielded analogous results (Docherty and McGrath 1979; de Jonge et al. 1983; Illes and Starke 1983; Pennefather 1983; Angus et al. 1984; Ledda and Mantelli 1984; Nilsson et al. 1985b; Stjärne and Åstrand 1985).

The "touchstones", then, have confirmed that presynaptic α_2-autoinhibition operates under appropriate conditions in vitro. However, three criticisms have been made.

On the basis of detailed studies in guinea-pig atria, Angus and Korner (1980) and Angus et al. (1984) concluded that the biophase concentration of noradrenaline remains too low for autoinhibition at physiologically low rates of impulse flow, and that autoinhibition is "a 'last resort' mechanism when the level of sympathetic efferent activity has become very high" (Angus et al. 1984). These studies have led to a closer look at the conditions for autoinhibition. There seems to be a consensus among many authors now, including the Australian group, that autoinhibition may not operate during the first few pulses in a train or when pulses are widely spaced, but does set in at higher frequencies; and, importantly, that the enhancement of release by α_2-antagonists is in fact due to interruption of autoinhibition and not to some undefined side effect (e.g., McCulloch et al. 1981; Story et al. 1981; Auch-Schwelk et al. 1983; Angus et al. 1984; Fuder et al. 1984; Heepe and Starke 1985; Nilsson et al. 1985b). A general restriction of autoinhibition to extreme rates of impulse traffic seems questionable. In brain cortex slices, for instance, autoinhibition operates at 1 Hz (Baumann and Koella 1980; Heepe and Starke 1985). This is not supraphysiologic for neurons that normally fire at 0.5–5 Hz. The normal irregular impulse pattern recorded from human vasoconstrictor fibers has recently been used to stimulate blood vessels in vitro (Nilsson et al. 1985a); a study of α_2-autoinhibition at such physiologic impulse series, with bursts separated by periods of quiescence, would be of interest.

The second criticism originated from studies on discrete events in the vas deferens. As mentioned above (Sect. 5.1), discrete events represent the release of transmitter by a single pulse from a single varicosity, and have revealed an intermittency of release. When discrete events elicited by trains of pulses were observed in the absence of any drug, the occurrence of one did *not* decrease the probability of the occurrence of a second one within the next few milliseconds to seconds, nor its quantal content. No sign of autoinhibition was detected (Blakeley et al. 1982; Cunnane and Stjärne 1984b). It seems that these findings, rather than questioning the reality of autoinhibition, indicate a mode of operation that escapes the method employed. In another study, Blakeley et al. (1984a) suggested that presynaptic α_2-autoreceptors may interfere with the process of facilitation in the course of a train of pulses, and that this particular mechanism "may go part way to explain why it is difficult to observe the action of such feedback on a stimulus to stimulus basis".

Other electrophysiologic studies have in general supported an endogenous autoinhibition (Blakeley et al. 1981; Nakamura et al. 1981; Illes and Starke 1983; Sneddon and Westfall 1984; Nörenberg 1986; see, however, Suzuki 1984; Blakeley et al. 1986).

The most far-reaching criticism has been advanced by Kalsner and his coworkers (see Kalsner 1985b). Chan and Kalsner (1979) originally doubted the very existence of presynaptic α-receptors. This position seems to have been abandoned (Kalsner 1982), but the view remains that the receptors are not sites of action of released noradrenaline and that, hence, there is no autoinhibition. Some aspects of this work have already been discussed (see Sects. 2, 3.1, 3.4). Three general points will follow.

1. Kalsner argues that autoinhibition should manifest itself as a fall of the per-pulse release of noradrenaline with increasing frequency. However, this would be so only if autoinhibition were the only determinant and if the basic phenomenon of frequency-dependent facilitation, for instance, did not exist (see Starke 1977; Alberts et al. 1981; Blakeley et al. 1984a, b). Facilitation seems to be weak in cerebrocortical noradrenaline axons. In brain cortex slices, the per-pulse release of noradrenaline indeed fell with increasing frequency (Montel et al. 1974; Baumann and Koella 1980; Cichini et al. 1986), and, interestingly, the decline was abolished by piperoxan (at least over a range of low frequencies; Baumann and Koella 1980).

2. It has been postulated that an enhancement by α-antagonists of postsynaptic effector responses must accompany the enhancement of noradrenaline release in order to make the hypothesis of an endogenous autoinhibition acceptable (Kalsner 1985b). But would not violation of this criterion argue against an intimate relationship between release and postsynaptic response (see Starke et al. 1972) rather than against the primarily presynaptic phenomenon of autoinhibition? Presynaptic modulation does have postsynaptic consequences, as will be discussed later (Sect. 6.4).

3. The contrast culminates in the tenet that α-adrenoceptor antagonists enhance the release of noradrenaline not by antagonism to noradrenaline but exclusively by other mechanisms (such as a TEA-like effect; see Kalsner and Quillan 1984; Heepe and Starke 1985). The two main arguments are that phenoxybenzamine increased the release of noradrenaline when only a single pulse was applied (and there should be little perineuronal noradrenaline; Kalsner 1979), and that the effect of the antagonists did not increase with the frequency of stimulation (e.g., Chan and Kalsner 1979; Kalsner 1981). The one-pulse experiments have been criticized because phenoxybenzamine was tested only at a

very high concentration (33 μmol/liter) with certainly a number of potential actions (Starke 1981b; Marshall 1983; McCulloch et al. 1985). Kalsner himself recently referred even to phenoxybenzamine 3 μmol/ liter as a concentration "too high to allow separation of multiple pre-synaptic effects" (Kalsner 1985b). Lower concentrations of α_2-antag-onists have rarely, if ever, increased the release of noradrenaline evoked by a single pulse (or by several widely spaced pulses; Story et al. 1981; Auch-Schwelk et al. 1983; Marshall 1983; Angus et al. 1984; Limberger and Starke 1984; Heepe and Starke 1985; in the study by Wakade and Wakade 1981, TEA was added and may have caused repetitive firing).

Even when we dismiss the one-pulse argument, however, the second argument remains. Studies were summarized above, in which the release-enhancing effect of α-antagonists did vary directly with the stimulation frequency. Yet attention was already drawn to the fact that in many – at least as many – studies, ever since the first report by Brown and Gil-lespie (1957), the opposite was found or a constant enhancement despite changes in frequency (e.g., Langer et al. 1975; Alberts et al. 1981; and the work of Kalsner; see also Sect. 5.2). Such contrary findings, when made repeatedly by several groups, urgently require reconciliation in the context of as much of the available knowledge as possible. This seems feasible here. The evidence presented above leaves little doubt that the release-enhancing effect of α_2-antagonists depends on, and *ceteris paribus* increases with, the biophase concentration of noradren-aline. Another large body of evidence indicates that presynaptic modula-tion generally declines with increasing stimulus "strength". These two influences militate against each other in the case of α_2-antagonists. The experimental conditions (and perhaps also the tissue selected) may be such that the influence of the stimulus strength prevails, and then the enhancement by α_2-antagonists will fall with increasing frequency (or increasing pulse width, or the addition of 4-AP, or use of very high K^+ for release). Yet I believe that (in most tissues at least) the ex-perimental conditions can also be chosen, with circumspection and patience, such that the influence of the noradrenaline concentration prevails, thus leading to increased enhancement with increasing frequency, so that also this last facet is added to the sum of findings that demonstrate endogenous presynaptic α_2-autoinhibition in vitro.

6.2 The Postsynaptic Response

Drugs with α_2-adrenoceptor affinity can influence the postsynaptic cells of noradrenergic synapses by a direct effect at postsynaptic α_2-receptors and by modulation of the release of noradrenaline. The second component will be best seen in simple peripheral neuroeffector junctions, especially where the postsynaptic receptors are predominantly α_1 or β. Like the degree of modulation of release, the degree of modulation of the postsynaptic response will depend on such factors as the frequency of stimulation and the presence or absence of uptake inhibitors. It will also depend on side effects not related to α_2-receptors. For instance, phentolamine antagonizes the heart rate effect of noradrenaline postsynaptically, and this will blunt any potentiation of cardiac sympathetic transmission (Langer et al. 1977; Drew 1980). An enhancement of postsynaptic responses will be small, moreover, when the response was already near-maximal before addition of the α-receptor antagonist.

The literature is replete with demonstrations of an inhibition, by selective α_2-adrenoceptor agonists, of sympathetic transmission to effectors with postsynaptic α_1- or β-receptors. Historically interesting examples are the inhibition by clonidine of the heart rate response, and the inhibition by xylazine of the nictitating membrane response, to sympathetic nerve stimulation in cats (Kobinger 1967; Kroneberg et al. 1967). Both were observed before the existence of presynaptic α-receptors was suspected.

There are also many demonstrations of an enhancement, by selective α_2-adrenoceptor antagonists, of sympathetic transmission to effectors with postsynaptic α_1- or β-receptors. Examples for tissues with postsynaptic α_1-receptors are dog blood vessels (Toda et al. 1984), the vasculature of the pithed rat (Yamaguchi and Kopin 1980; Zukowska-Grojec et al. 1983), the rat mesenteric vascular bed (Eikenburg 1984; Nilsson et al. 1985b) and vas deferens (Vizi et al. 1973; Brown et al. 1979; French and Scott 1983; Pennefather 1983), the rabbit aorta (Docherty and Starke 1982), pulmonary artery (Starke et al. 1975b; Constantine et al. 1978; Lattimer et al. 1984), and ileocolic artery (von Kügelgen and Starke 1985), and the rabbit hindlimb (Madjar et al. 1980). An even greater number of studies have shown an increase in cardiac responses to sympathetic nerve impulses. Examples are experiments on isolated preparations of the heart of the cat (Ilhan et al. 1976), rat (Enero 1984), guinea pig (Langer et al. 1977; McCulloch et al. 1981; Story et al. 1981; Belleau et al. 1982; Angus et al. 1984; Ledda and Mantelli 1984) and rabbit (Carr and Fozard 1981) as well as on the heart in situ of the dog (Constantine et al. 1978; Cavero et al. 1979; Shepperson et al. 1981; Dabiré et al. 1981 with a cautionary comment; Heyndrickx et al. 1984; Saeed et al. 1985; Yorikane et al. 1986), cat (Abdel Rahman and Sharabi 1981), and pithed rat (Algate and Waterfall 1978; Docherty and McGrath 1979; Cavero et al. 1980; Drew 1980; Demichel et al. 1982; de Jonge et al. 1983). There are exceptions in which α-receptor antagonists failed to increase cardiac responses. Some are open to explanation. The tachycardia elicited in rat atria by 5 pulses delivered at 10 Hz was not increased by phentolamine (5 μmol/liter) (Idowu and Zar 1977); yet this pulse series may have been too brief for α-autoinhibition to develop, and the rather high concentration of phentolamine may have counteracted released noradrenaline postsynaptically (see above). Phentolamine did not increase the sympathetically evoked cardioacceleration in dogs in a study by Antonaccio et al. (1974); however, clonidine also failed to depress the cardiac response in these experiments, an observation which contrasts with many other reports. On the whole, the evidence for an increase by α_2-antagonists of sympathetic neuroeffector transmission seems overwhelming – with the important implication that *autoinhibition by released noradrenaline, although primarily a presynaptic event, secondarily attenuates the postsynaptic response.*

Notwithstanding this qualitative statement, it must be admitted that quantitatively, the increase in postsynaptic responses is often small (see the divergent conclusions of Drew 1980 and de Jonge et al. 1983). α_2-Adrenoceptor antagonists in general enhance the release of noradrenaline at best fivefold. This may not mean very much in terms of the effector cell response. Yet one should perhaps not conclude that α-autoinhibition in the end is negligible. Fivefold variations in transmitter release and, say, 30% variations in postsynaptic responses may not decide between life and death, but they might well contribute to deciding between disease and well-being.

6.3 α_2-Autoreceptors In Vivo: Biochemical Evidence

Experiments indicating the operation of presynaptic α_2-autoreceptors in vivo have occasionally been mentioned above. This section summarizes the biochemical evidence available. The evidence is mainly of two kinds: measurement of the venous overflow of noradrenaline from individual tissues, and measurement of the plasma noradrenaline level or the total body "noradrenaline spillover rate". Both the plasma level and the spillover rate reflect the total peripheral release of noradrenaline. The spillover rate is the better index; it is the rate at which endogenous noradrenaline enters the plasma, and changes in the plasma noradrenaline clearance are taken into account in its calculation (Esler et al. 1984). The discussion will be limited to peripheral sympathetically innervated tissues. Observations on pithed animals with electrically stimulated sympathetic outflow will be included.

The study at the very source of the autoreceptor hypothesis must be quoted here as the first overflow study in vivo: phenoxybenzamine and dibenamine markedly increased the electrically evoked overflow of noradrenaline from the in situ blood-perfused cat spleen (Brown and Gillespie 1957). More recent investigations have mostly been carried out in dogs. α_2-Adrenoceptor agonists depress, and α_2-antagonists enhance, the coronary overflow of noradrenaline elicited by sympathetic nerve stimulation in the heart of anesthetized and spinal dogs (Yamaguchi et al. 1977; Cavero et al. 1979; Cousineau et al. 1984; Saeed et al. 1985; Yorikane et al. 1986; see also Heyndrickx et al. 1984). The findings mirror the postsynaptic response changes mentioned above. Rauwolscine and phenoxybenzamine, but not prazosin, also increased the evoked overflow of noradrenaline from the dog hindleg and gracilis muscle (Elsner et al. 1984; Kahan et al. 1984), while clonidine caused inhibition and yohimbine enhancement in the liver (Yamaguchi 1982). In anesthetized pigs, phentolamine increased the evoked overflow of noradrenaline (and its cotransmitter neuropeptide Y) from the spleen (Lundberg et al. 1986), and in unanesthetized, freely moving rats yohimbine increased the evoked overflow from the portal vein (Remie and Zaagsma 1986). These studies are all the more interesting since they demonstrate modulation of the release of endogenous rather than previously incorporated radiolabeled noradrenaline. The study in the pig spleen nicely shows, moreover, how in vivo as well as in vitro (Sect. 6.1) transmitter release (measured as overflow of neuropeptide Y) and the effect of phentolamine depend on the biophase concentration of noradrenaline (Lundberg et al. 1986).

It is well known that clonidine-like α_2-agonists lower the plasma noradrenaline level and the noradrenaline spillover rate (see Esler et al. 1984). The effect is generally assumed to be central in origin. Yet α_2-agonists can decrease the total peripheral release of noradrenaline via peripheral receptors as well. Clonidine, α-methylnoradrenaline, and adrenaline reduced the noradrenaline spillover rate in pithed rabbits with electrically stimulated sympathetic outflow, an effect blocked by yohimbine and phenoxybenzamine (Majewski et al. 1983c, 1985). A central action is excluded in these preparations. A central site also seems unlikely in experiments in which α-methylnoradrenaline (but not an equihypertensive dose of phenylephrine) decreased the noradrenaline spillover rate in conscious rabbits, an effect again blocked by yohimbine (Majewski et al. 1983b); α-methylnoradrenaline will not easily cross the blood-brain barrier. A peripheral component has been suggested even in the case of clonidine in animals with an intact central nervous system: in anesthetized rats, the fall in plasma noradrenaline following intravenous clonidine was prevented by intravenous idazoxan but not by intracerebroventricular idazoxan (Brown and Harland 1984). α-Methyl-noradrenaline failed to decrease the noradrenaline spillover rate in humans, and this seemed to cast doubt on the operation of α_2-auto-receptors in man (FitzGerald et al. 1981). Possibly, however, the dose was too low. More recent experiments from the same laboratory have led to some revision of the original view. Clonidine, guanfacine, and adrenaline all reduced the plasma noradrenaline level in man, apparently by activation of peripheral presynaptic α_2-autoreceptors (Murphy et al. 1984; Brown et al. 1985).

Finally, does endogenous noradrenaline share this effect and is there, hence, a component of autoinhibition in the total peripheral release of noradrenaline that might be uncovered by antagonists? In pithed rats with electrically stimulated sympathetic outflow, yohimbine, piperoxan, tolazoline, phentolamine, and phenoxybenzamine all increased the plasma noradrenaline level. Moreover, the pressor response to stimulation was enhanced by low doses of the α_2-selective antagonists yohimbine and piperoxan exactly as expected from the increase in noradrenaline release (Yamaguchi and Kopin 1980; Zukowska-Grojec et al. 1983). This elaborate series of findings displays clearly presynaptic autoinhibi-tion and its postsynaptic sequelae in a whole animal preparation.

A stepwise approach to the normal animal is shown in Fig. 6. The α_1-selective antagonists prazosin and corynanthine and the α_2-selective antagonists yohimbine and rauwolscine were administered to pithed rabbits with stimulation of the sympathetic outflow (Fig. 6A), anes-thetized rabbits (Fig. 6B), and conscious rabbits (Fig. 6C). In the pithed animals, all antagonists decreased mean arterial pressure (by blockade of postsynaptic α_1- and possibly α_2-adrenoceptors), but only the α_2-selective antagonists increased the noradrenaline spillover rate. In anes-thetized rabbits, yohimbine did not change blood pressure whereas rauwolscine decreased it. As a check for baroreceptor-mediated reflex adjustments in the release of noradrenaline, rauwolscine was compared with the vasodilators hydralazine and sodium nitroprusside, which were given at equihypotensive doses. Only yohimbine and rauwolscine increased the spillover of noradrenaline. In conscious rabbits, finally, corynanthine did not change blood pressure but yohimbine and rauwolscine slightly increased it. Again, only the α_2-selective alkaloids enhanced the release of noradrenaline. With the approach to the con-dition of a normal animal, effects mediated by the central nervous system probably become more and more prominent, and central sympatho-excitation may be responsible for the large increase in

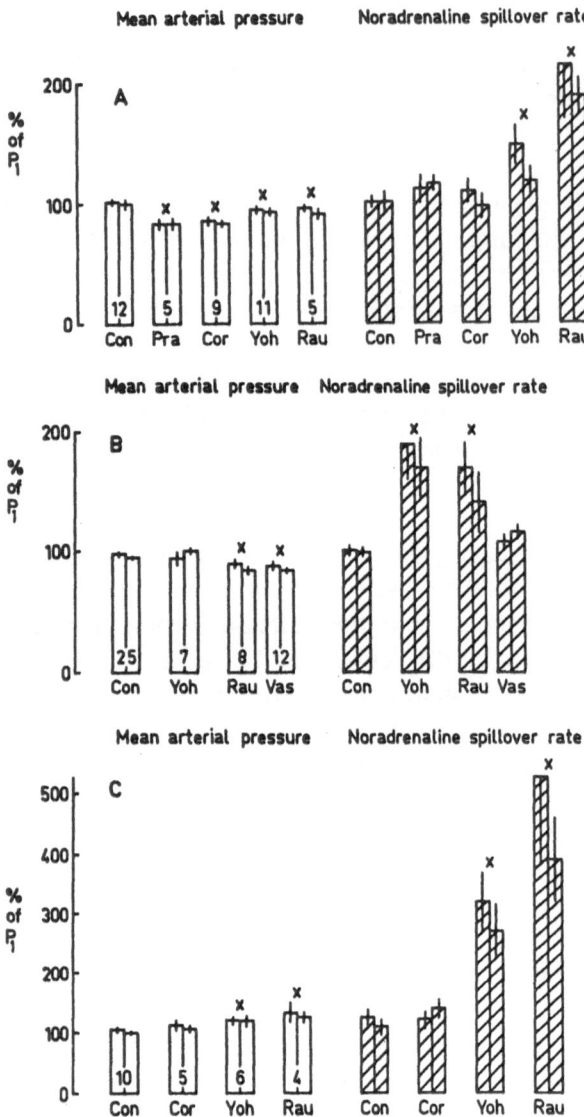

Fig. 6 A–C. Effects of prazosin (*Pra*, 0.1 mg/kg), corynanthine (*Cor*, 1 mg/kg), yohimbine (*Yoh*, 1 mg/kg), and rauwolscine (*Rau*, 1 mg/kg) on mean arterial pressure and the noradrenaline spillover rate in pithed rabbits with electrically stimulated sympathetic outflow (**A**), anesthetized rabbits (**B**), and conscious rabbits (**C**). All parameters were measured three times, before (P_1) and 20–30 min (P_2) and 40–60 min (P_3) after intravenous injection of the drugs. For comparison with rauwolscine in anesthetized rabbits, the vasodilator drugs hydralazine and sodium nitroprusside were administered after P_1 in equihypotensive doses (*Vas;* pooled results). The two adjoining columns represent P_2 and P_3, expressed as a percentage of P_1 values. Number of experiments in the columns. Significant differences from control (*Con*): × P < 0.05. Modified from Majewski et al. (1983a–c)

noradrenaline release that yohimbine and rauwolscine produced in conscious rabbits. Yet the consistency of the effects of the α_2-selective compounds suggests that an interruption of peripheral presynaptic α_2-autoinhibition — the only mechanism in the pithed rabbits — also contributed to the increase in the anesthetized and conscious animals (see also Graham et al. 1980; Brown and Harland 1984; the apparently negative result with phentolamine reported by Hamilton et al. 1982 may be due to the partial α_2-agonist character of this compound).

For obvious reasons, there is less evidence for an endogenous auto-inhibition of peripheral noradrenaline release in vivo in man. Idazoxan has been shown to cause a slight increase in blood pressure and an increase in the plasma noradrenaline concentration (Elliott et al. 1984; Brown et al. 1985). Both groups conclude that the observation is consistent with the interruption of an ongoing physiologic autoinhibition, and Brown et al. (1985), on the basis of previous animal experiments, raise this possibility to the rank of probability.

7 Outlook

What is the basic significance of α-autoinhibition? Does it play "a major physiological role in noradrenergic neurotransmission" (Langer 1977)? Is its purpose "a reduction of effects produced by stray or nonintended stimuli of low frequency" (von Euler 1979)? Is it "a 'last resort' mechanism" (Angus et al. 1984)? Does it, if indeed it exists, have "an impact on the dimensions of the effector response which can only be described as trivial" (Kalsner 1983)? Does it belong to those presynaptic receptor mechanisms that are functionless "vestiges of evolution that continue to exist because they do us no harm" (Starke 1981b)? I see no reason to depart greatly from the skeptical optimism expressed previously, i.e., "that the advantage provided by the pre- and trans-synaptic feedbacks is far from being clear. Anyway, even if the sense is not clear we have to accept the fact that these mechanisms work ...; prostaglandin-, α- and/or β-receptor-mediated feedback mechanisms contribute to the regulation of noradrenergic transmission" (Starke 1977). In some tissues and situations, this may mean little; in other tissues and situations, presynaptic autoinhibition and its postsynaptic consequences may be substantial.

There are important aspects that have not been discussed here. Endogenous α_2-autoinhibition seems to operate in vivo in the periphery (see Sect. 6.3). Does it also operate in vivo in the central nervous system (Dietl et al. 1981; Tepper et al. 1985; Carter et al. 1986; L'Heureux et al. 1986)? Does it have pathophysiologic implications, as has been suggested for hypertension (see Westfall et al. 1985; Kubo et al. 1986), or therapeutic implications, perhaps in the effect of antidepressants or clonidine-like antihypertensives (see Keith et al. 1986 and Jonkman et al. 1985 for recent studies)? Do α_2-autoreceptors change with age (Docherty 1986)? Is adrenal medullary hormone release also modulated by α_2-adrenoceptor ligands? This seemed to be an interesting possibility for some time but has now been made rather unlikely (see Powis and Baker 1986; Sharma et al. 1986). Do presynaptic α_2-autoreceptors inhibit the synthesis of noradrenaline from tyrosine, independently of the inhibition of release (Birch and Fillenz 1985)? This would be contrary to a previous suggestion (Grabowska and Andén 1976) but would be a parallel to dopaminergic terminal axons for which independent autoreceptor-mediated inhibition of transmitter release and transmitter synthesis have long been postulated. Since many terminals possess numerous receptors, surely with overlapping postreceptor mechanisms, will the receptors interact? There is growing evidence for this possibility (α_2- and β-receptors: see Johnston and Majewski 1986; α_2-receptors and muscarine receptors: Loiacono et al. 1985; α_2-receptors and opioid receptors: Hölting and Starke 1986; Limberger et al. 1986; Ramme et al. 1986; Schoffelmeer et al. 1986a).

Finally, cotransmission opens up a new perspective. An axon releasing both noradrenaline and ATP, for instance, may transmit information through postsynaptic adrenoceptors and purine P_2-receptors. In addition, it may possess a second kind of presynaptic autoinhibition. ATP is rapidly hydrolyzed to adenosine, which may then inhibit transmitter release through presynaptic purine P_1-receptors. The dynamics of synaptic systems consisting of two transmitters and several pre- and postsynaptic transmitter receptors are now beginning to be studied (noradrenaline-ATP: Sneddon and Westfall 1984; Sneddon et al. 1984; Stjärne and Åstrand 1985; von Kügelgen and Starke 1985; noradrenaline-neuropeptide Y: Lundberg et al. 1986; Pernow et al. 1986; Stjärne et al. 1986b). α_2-Autoinhibition may be a powerful mechanism in a noradrenaline-ATP synapse in which ATP is an important signal for the postsynaptic cell, as for instance in the ileocolic artery of the rabbit. When the axons within the arteries were stimulated by 100 pulses at 5 Hz, yohimbine increased the ensuing vasoconstriction no less than threefold, without any change in the response to exogenous noradrenaline, thus indicating a marked increase in the release of noradrenaline and presumably ATP, and uncovering a presynaptic autoinhibition with pronounced postsynaptic results (von Kügelgen and Starke 1985).

The α_2-autoreceptor hypothesis has both solved and raised problems. It promises further *quaestiones disputatae* and clarifying answers for the future.

References

Abdel Rahman ARA, Sharabi FM (1981) Presynaptic alpha receptors in relation to the cardiovascular effect of yohimbine in the anesthetized cat. Arch int Pharmacodyn Ther 252:229–240

Aghajanian GK, Wang YY (1986) Pertussis toxin blocks the outward currents evoked by opiate and α_2-agonists in locus coeruleus neurons. Brain Res 371:390–394

Alabaster VA, Keir RF, Peters CJ (1985) Comparison of activity of alpha-adrenoceptor agonists and antagonists in dog and rabbit saphenous vein. Naunyn-Schmiedeberg's Arch Pharmacol 330:33–36

Alberts P, Stjärne L (1982) Facilitation, and muscarinic and α-adrenergic inhibition of the secretion of ^{3}H-acetylcholine and ^{3}H-noradrenaline from guinea-pig ileum myenteric nerve terminals. Acta Physiol Scand 116:83–92

Alberts P, Bartfai T, Stjärne L (1981) Site(s) and ionic basis of α-autoinhibition and facilitation of [^{3}H]noradrenaline secretion in guinea-pig vas deferens. J Physiol (Lond) 312:297–334

Alberts P, Ögren VR, Sellström ÅI (1985) Role of adenosine $3',5'$-cyclic monophosphate in adrenoceptor-mediated control of ^{3}H-noradrenaline secretion in guinea-pig ileum myenteric nerve terminals. Naunyn-Schmiedeberg's Arch Pharmacol 330:114–120

Algate DR, Waterfall JF (1978) Action of indoramin on pre- and postsynaptic α-adrenoceptors in pithed rats. J Pharm Pharmacol 30:651–652

Allgaier C, Hertting G (1986a) Involvement of protein kinase C in the modulation of noradrenaline release via α_2-adrenoceptors. Naunyn-Schmiedeberg's Arch Pharmacol 332:R78

Allgaier C, Hertting G (1986b) Polymyxin B, a selective inhibitor of protein kinase C, diminishes the release of noradrenaline and the enhancement of release caused by phorbol 12,13-dibutyrate. Naunyn-Schmiedeberg's Arch Pharmacol 334:218–221

Allgaier C, Feuerstein TJ, Jackisch R, Hertting G (1985) Islet-activating protein (pertussis toxin) diminishes α_2-adrenoceptor mediated effects on noradrenaline release. Naunyn-Schmiedeberg's Arch Pharmacol 331:235–239

Allgaier C, Feuerstein TJ, Hertting G (1986a) N-ethylmaleimide (NEM) diminishes α_2-adrenoceptor mediated effects on noradrenaline release. Naunyn-Schmiedeberg's Arch Pharmacol 333:104–109

Allgaier C, von Kügelgen O, Hertting G (1986b) Enhancement of noradrenaline release by 12-0-tetradecanoyl phorbol-13-acetate, an activator of protein kinase C. Eur J Pharmacol 129:389–392

Alonso FG, Ceña V, García AG, Kirpekar SM, Sánchez-García P (1982) Presence and axonal transport of cholinoceptor, but not adrenoceptor sites on a cat noradrenergic neurone. J Physiol (Lond) 333:595–618

Andrade R, Aghajanian GK (1985) Opiate- and α_2-adrenoceptor-induced hyperpolarizations of locus ceruleus neurons in brain slices: reversal by cyclic adenosine $3':5'$-monophosphate analogues. J Neurosci 5:2359–2364

Angus JA, Korner PI (1980) Evidence against presynaptic α-adrenoceptor modulation of cardiac sympathetic transmission. Nature 286:288–291

Angus JA, Bobik A, Jackman GP, Kopin IJ, Korner PI (1984) Role of auto-inhibitory feed-back in cardiac sympathetic transmission assessed by simultaneous measurements of changes in ^{3}H-efflux and atrial rate in guinea-pig atrium. Br J Pharmacol 81:201–214

Antonaccio MJ, Halley J, Kerwin L (1974) Functional significance of α-stimulation and α-blockade on responses to cardiac nerve stimulation in anesthetized dogs. Life Sci 15:765–777

Asakura M, Tsukamoto T, Imafuku J, Matsui H, Ino M, Hasegawa K (1985) Quantitative analysis of rat brain α_2-receptors discriminated by [^{3}H] clonidine and [^{3}H] rauwolscine. Eur J Pharmacol 106:141–147

Atlas D, Burstein Y (1984) Isolation and partial purification of a clonidine-displacing endogenous brain substance. Eur J Biochem 144:287–293

Auch-Schwelk W, Starke K, Steppeler A (1983) Experimental conditions required for the enhancement by α-adrenoceptor antagonists of noradrenaline release in the rabbit ear artery. Br J Pharmacol 78:543–551

Baker DJ, Drew GM, Hilditch A (1984) Presynaptic α-adrenoceptors: do exogenous and neuronally released noradrenaline act at different sites? Br J Pharmacol 81:457–464

Baumann PA, Koella WP (1980) Feedback control of noradrenaline release as a function of noradrenaline concentration in the synaptic cleft in cortical slices of the rat. Brain Res 189:437–448

Belis JA, Colby JE, Westfall DP (1982) Effects of α-adrenoceptor agents on norepinephrine release from vas deferens of several species including man. Eur J Pharmacol 78:487–490

Belleau B, Benfey BG, Melchiorre C (1982) Presynaptic effect of clonidine antagonized by the tetramine disulphide, benextramine. Br J Pharmacol 75:617–621

Bevan JA, Tayo FM, Rowan RA, Bevan RD (1984) Presynaptic α-receptor control of adrenergic transmitter release in blood vessels. Fed Proc 43:1365–1370

Birch PJ, Fillenz M (1985) Stimulation of noradrenaline synthesis by the calcium ionophore A23187 and its modulation by presynaptic receptors. Neurosci Lett 62:187–192

Blakeley AGH, Cunnane TC (1979) The packeted release of transmitter from the sympathetic nerves of the guinea-pig vas deferens: an electrophysiological study. J Physiol (Lond) 296:85–96

Blakeley AGH, Cunnane TC, Petersen SA (1981) An electropharmacological analysis of the effects of some drugs on neuromuscular transmission in the vas deferens of the guinea-pig. J Auton Pharmacol 1:367–375

Blakeley AGH, Cunnane TC, Petersen SA (1982) Local regulation of transmitter release from rodent sympathetic nerve terminals? J Physiol (Lond) 325:93–109

Blakeley AGH, Cunnane TC, Maskell T, Mathie A, Petersen SA (1984a) α-Adrenoreceptors and facilitation at a sympathetic neuroeffector junction. J Auton Pharmacol 4:53–58

Blakeley AGH, Mathie A, Petersen SA (1984b) Facilitation at single release sites of a sympathetic neuroeffector junction in the mouse. J Physiol (Lond) 349:57–71

Blakeley AGH, Mathie A, Petersen SA (1986) Interactions between the effects of yohimbine, clonidine and $[Ca]_o$ on the electrical response of the mouse vas deferens. Br J Pharmacol 88:807–814

Bokoch GM, Smigel M, Higashijima T, Gilman AG (1985) Guanine nucleotide-binding regulatory proteins as information transducers. In: Lefkowitz RJ, Lindenlaub E (eds) Adrenergic receptors: molecular properties and therapeutic implications. Schattauer, Stuttgart, pp 135–156

Bonanno G, Raiteri M (1987) Interaction between 5-HT uptake inhibition and activation of 5-HT autoreceptors by exogenous agonists in rat cerebral cortex slices and synaptosomes. Naunyn-Schmiedeberg's Arch Pharmacol (in press)

Bond RA, Charlton KG, Clarke DE (1986) Responses to norepinephrine resistant to inhibition by alpha and beta adrenoceptor antagonists. J Pharmacol Exp Ther 236:408–415

Bousquet P, Feldman J, Atlas D (1986) An endogenous, non-catecholamine clonidine antagonist increases mean arterial blood pressure. Eur J Pharmacol 124:167–170

Bradberry CW, Adams RN (1986) $α_2$-Receptor control over release of noradrenaline in rat thalamus. Eur J Pharmacol 129:175–180

Bradley L, Doggrell SA (1983) Effects of prazosin, phentolamine and yohimbine on noradrenergic transmission in the rat right ventricle in vitro. J Auton Pharmacol 3:27–36

Brown CM, McGrath JC, Summers RJ (1979) The effects of α-adrenoceptor agonists and antagonists on responses of transmurally stimulated prostatic and epididymal portions of the isolated vas deferens of the rat. Br J Pharmacol 66: 553–564

Brown DA, Caulfield MP (1979) Hyperpolarizing 'α₂'-adrenoceptors in rat sympathetic ganglia. Br J Pharmacol 65:435–445

Brown GL, Gillespie JS (1957) The output of sympathetic transmitter from the spleen of the cat. J Physiol (Lond) 138:81–102

Brown MJ, Harland D (1984) Evidence for a peripheral component in the sympatholytic effect of clonidine in rats. Br J Pharmacol 83:657–665

Brown MJ, Struthers AD, Burrin JM, Di Silvio L, Brown DC (1985) The physiological and pharmacological role of presynaptic α- and β-adrenoceptors in man. Br J clin Pharmacol 20:649–658

Browning MD, Huganir R, Greengard P (1985) Protein phosphorylation and neuronal function. J Neurochem 45:11–23

Bug W, Williams JT, North RA (1986) Membrane potential measured during potassium-evoked release of noradrenaline from rat brain neurons: effects of normorphine. J Neurochem 47:652–655

Canfield DR, Dunlap K (1984) Pharmacological characterization of amine receptors on embryonic chick sensory neurones. Br J Pharmacol 82:557–563

Carr SR, Fozard JR (1981) Lack of modulation by presynaptic α₂-adrenoceptors of adrenergic transmitter release evoked by activation of 5-hydroxytryptamine and nicotine receptors. Eur J Pharmacol 72:27–34

Carter CJ, Dennis T, L'Heureux R, Scatton B (1986) Studies of the effects of idazoxan on cortical noradrenaline release in vivo, measured by transcortical dialysis. Br J Pharmacol 88:307P

Cavero I, Dennis T, Lefèvre-Borg F, Perrot P, Roach AG, Scatton B (1979) Effects of clonidine, prazosin and phentolamine on heart rate and coronary sinus catecholamine concentration during cardioaccelerator nerve stimulation in spinal dogs. Br J Pharmacol 67:283–292

Cavero I, Gomeni R, Lefèvre-Borg F, Roach AG (1980) Comparison of mianserin with desipramine, maprotiline and phentolamine on cardiac presynaptic and vascular postsynaptic α-adrenoceptors and noradrenaline reuptake in pithed normotensive rats. Br J Pharmacol 68:321–332

Celuch SM, Dubocovich ML, Langer SZ (1978) Stimulation of presynaptic β-adrenoceptors enhances [³H]-noradrenaline release during nerve stimulation in the perfused cat spleen. Br J Pharmacol 63:97–109

Ceña V, García AG, Khoyi MA, Salaices M, Sanchez-García P (1985) Effect of the dihydropyridine Bay K 8644 on the release of [³H]-noradrenaline from the rat isolated vas deferens. Br J Pharmacol 85:691–696

Chan CC, Kalsner S (1979) An examination of the negative feedback function of presynaptic adrenoceptors in a vascular tissue. Br J Pharmacol 67:401–407

Cichini G, Lassmann H, Placheta P, Singer EA (1986) Effects of clonidine on the stimulation-evoked release of ³H-noradrenaline from superfused rat brain slices as a function of the biophase concentration. Naunyn-Schmiedeberg's Arch Pharmacol 333:36–42

Cohen RA, Shepherd JT, Vanhoutte PM (1983) Prejunctional and postjunctional actions of endogenous norepinephrine at the sympathetic neuroeffector junction in canine coronary arteries. Circul Res 52:16–25

Cole AE, Shinnick-Gallagher P (1981) Comparison of the receptors mediating the catecholamine hyperpolarization and slow inhibitory postsynaptic potential in sympathetic ganglia. J Pharmacol Exp Ther 217:440–444

Constantine JW, Weeks RA, McShane WK (1978) Prazosin and presynaptic α-receptors in the cardioaccelerator nerve of the dog. Eur J Pharmacol 50:51–60

Constantine JW, Gunnell D, Weeks RA (1980) α₁- and α₂-vascular adrenoceptors in the dog. Eur J Pharmacol 66:281–286

Cousineau D, Goresky CA, Bach GG, Rose CP (1984) Effect of β-adrenergic blockade on in vivo norepinephrine release in canine heart. Am J Physiol 246:H283–H292

Cubeddu L, Weiner N (1975a) Nerve stimulation-mediated overflow of norepinephrine and dopamine-β-hydroxylase. III. Effects of norepinephrine depletion on the alpha presynaptic regulation of release. J Pharmacol Exp Ther 192:1–14

Cubeddu L, Weiner N (1975b) Release of norepinephrine and dopamine-β-hydroxylase by nerve stimulation. V. Enhanced release associated with a granular effect of a benzoquinolizine derivative with reserpine-like properties. J Pharmacol Exp Ther 193:757–774

Cubeddu L, Barnes E, Weiner N (1975) Release of norepinephrine and dopamine-β-hydroxylase by nerve stimulation. IV. An evaluation of a role for cyclic adenosine monophosphate. J Pharmacol Exp Ther 193:105–127

Cunnane TC, Stjärne L (1984a) Frequency dependent intermittency and ionic basis of impulse conduction in postganglionic sympathetic fibres of guinea-pig vas deferens. Neuroscience 11:211–229

Cunnane TC, Stjärne L (1984b) Transmitter secretion from individual varicosities of guinea-pig and mouse vas deferens: highly intermittent and monoquantal. Neuroscience 13:1–20

Dabiré H, Mouillé P, Andréjak M, Fournier B, Schmitt H (1981) Pre- and postsynaptic α-adrenoceptor blockade by (imidazolinyl-2)-2-benzodioxane 1–4 (170 150): antagonistic action on the central effects of clonidine. Arch int Pharmacodyn Ther 254:252–270

Dart AM, Dietz R, Hieronymus K, Kübler W, Mayer E, Schömig A, Strasser R (1984) Effects of α- and β-adrenoceptor blockade on the neurally evoked overflow of endogenous noradrenaline from the rat isolated heart. Br J Pharmacol 81:475–478

De Jonge A, Santing PN, Timmermans PBMWM, van Zwieten PA (1983) Effect of age on the prejunctional α-adrenoceptor-mediated feedback in the heart of spontaneously hypertensive rats and normotensive Wistar Kyoto rats. Naunyn-Schmiedeberg's Arch Pharmacol 323:33–36

De Jonge A, van den Berg G, Qian JQ, Wilffert B, Thoolen MJMC, Timmermans PBMWM, van Zwieten PA (1986) Inhibitory effect of alpha-1 adrenoceptor stimulation on cardiac sympathetic neurotransmission in pithed normotensive rats. J Pharmacol Exp Ther 236:500–504

De Langen CDJ, Mulder AH (1980) On the role of calcium ions in the presynaptic alpha-receptor mediated inhibition of [^{3}H]noradrenaline release from rat brain cortex synaptosomes. Brain Res 185:399–408

De Langen CDJ, Hogenboom F, Mulder AH (1979) Presynaptic noradrenergic α-receptors and modulation of ^{3}H-noradrenaline release from rat brain synaptosomes. Eur J Pharmacol 60:79–89

Demichel P, Gomond P, Roquebert J (1982) α-Adrenoceptor blocking properties of raubasine in pithed rats. Br J Pharmacol 77:449–454

Diamant S, Atlas D (1986) An endogenous brain substance, CDS (clonidine-displacing substance), inhibits the twitch response of rat vas deferens. Biochem Biophys Res Comm 134:184–190

Dietl H, Sinha JN, Philippu A (1981) Presynaptic regulation of the release of catecholamines in the cat hypothalamus. Brain Res 208:213–218

Digges KG, Summers RJ (1983) Effects of yohimbine stereoisomers on contractions of rat aortic strips produced by agonists with different selectivity for α_1- and α_2-adrenoceptors. Eur J Pharmacol 96:95–99

Dismukes RK, Mulder AH (1976) Cyclic AMP and α-receptor-mediated modulation of noradrenaline release from rat brain slices. Eur J Pharmacol 39:383–388

Dismukes K, De Boer AA, Mulder AH (1977) On the mechanism of alpha-receptor mediated modulation of ^{3}H-noradrenaline release from slices of rat brain neocortex. Naunyn-Schmiedeberg's Arch Pharmacol 299:115–122

Dixon WR, Mosimann WF, Weiner N (1979) The role of presynaptic feedback mechanisms in regulation of norepinephrine release by nerve stimulation. J Pharmacol Exp Ther 209:196–204

Docherty JR (1984) An investigation of presynaptic α-adrenoceptor subtypes in the pithed rat heart and in the rat isolated vas deferens. Br J Pharmacol 82: 15–23

Docherty JR (1986) Aging and the cardiovascular system. J Auton Pharmacol 6:77–84

Docherty JR, Hyland L (1985) No evidence for differences between pre- and post-junctional α_2-adrenoceptors. Br J Pharmacol 86:335–339

Docherty JR, McGrath JC (1979) An analysis of some factors influencing α-adreno-ceptor feed-back at the sympathetic junction in the rat heart. Br J Pharmacol 66:55–63

Docherty JR, Starke K (1982) An examination of the pre- and postsynaptic α-adrenoceptors involved in neuroeffector transmission in rabbit aorta and portal vein. Br J Pharmacol 76:327–335

Doggrell SA, Vincent L (1980) Accumulation and overflow of ^{3}H following incubation of the guinea-pig gall bladder with [^{3}H]-noradrenaline. Br J Pharmacol 71:557–567

Dooley DJ, Bittiger H, Hauser KL, Bischoff SF, Waldmeier PC (1983) Alteration of central alpha$_2$- and beta-adrenergic receptors in the rat after DSP-4, a selective noradrenergic neurotoxin. Neuroscience 9:889–898

Doxey JC, Lane AC, Roach AG, Virdee NK (1984) Comparison of the α-adrenoceptor antagonist profiles of idazoxan (RX 781094), yohimbine, rauwolscine and corynanthine. Naunyn-Schmiedeberg's Arch Pharmacol 325:136–144

Drew GM (1977) Pharmacological characterisation of the presynaptic α-adreno-ceptor in the rat vas deferens. Eur J Pharmacol 42:123–130

Drew GM (1978a) The effect of different calcium concentrations on the inhibitory effect of presynaptic α-adrenoceptors in the rat vas deferens. Br J Pharmacol 63:417–419

Drew GM (1978b) Pharmacological characterization of the presynaptic α-adreno-ceptors regulating cholinergic activity in the guinea-pig ileum. Br J Pharmacol 64:293–300

Drew GM (1980) Presynaptic modulation of heart rate responses to cardiac nerve stimulation in pithed rats. J Cardiovasc Pharmacol 2:843–856

Duckles SP (1982) Modulation of endogenous noradrenaline release by prejunctional α-adrenoreceptors: comparison of a cerebral and peripheral artery. J Auton Pharmacol 2:71–77

Dunlap K, Fischbach GD (1981) Neurotransmitters decrease the calcium conductance activated by depolarization of embryonic chick sensory neurones. J Physiol (Lond) 317:519–535

Dzielak DJ, Thureson-Klein Å, Klein RL (1983) Local modulation of neurotransmitter release in bovine splenic vein. Blood Vess 20:122–134

Ebstein RP, Seamon K, Creveling CR, Daly JW (1982) Release of norepinephrine from brain vesicular preparations: effects of an adenylate cyclase activator, forskolin, and a phosphodiesterase inhibitor. Cell Mol Neurobiol 2:179–192

Eikenburg DC (1984) Functional characterization of the pre- and postjunctional α-adrenoceptors in the in situ perfused rat mesenteric vascular bed. Eur J Pharmacol 105:161–165

Elliott HL, Jones CR, Vincent J, Lawrie CB, Reid JL (1984) The alpha adrenoceptor antagonist properties of idazoxan in normal subjects. Clin Pharmacol Ther 36:190–196

Elsner D, Saeed M, Sommer O, Holtz J, Bassenge E (1984) Sympathetic vaso-constriction sensitive to α_2-adrenergic receptor blockade. Hypertension 6: 915–925

Enero MA (1984) Influence of neuronal uptake on the presynaptic α-adrenergic modulation of noradrenaline release. Naunyn-Schmiedeberg's Arch Pharmacol 328:38–40

Enero MA, Langer SZ (1973) Influence of reserpine-induced depletion of noradrenaline on the negative feed-back mechanism for transmitter release during nerve stimulation. Br J Pharmacol 49:214–225

Ennis C, Lattimer N (1984) Presynaptic agonist effect of phentolamine in the rabbit vas deferens and rat cerebral cortex. J Pharm Pharmacol 36:753–757

Ercan ZS (1983) Prejunctional alpha-adrenoceptor mediated prostaglandin releasing effect of clonidine in the isolated perfused rabbit kidney. Arch int Pharmacodyn Ther 265:138–149

Esler M, Jennings G, Korner P, Blombery P, Sacharias N, Leonard P (1984) Measurement of total and organ-specific norepinephrine kinetics in humans. Am J Physiol 247:E21–E28

Farah MB, Langer SZ (1974) Protection by phentolamine against the effects of phenoxybenzamine on transmitter release elicited by nerve stimulation in the perfused cat heart. Br J Pharmacol 52:549–557

Feuerstein TJ, Hertting G, Jackisch R (1985) Endogenous noradrenaline as modulator of hippocampal serotonin (5-HT)-release. Naunyn-Schmiedeberg's Arch Pharmacol 329:216–221

Filinger EJ, Langer SZ, Perec CJ, Stefano FJE (1978) Evidence for the presynaptic location of the alpha-adrenoceptors which regulate noradrenaline release in the rat submaxillary gland. Naunyn-Schmiedeberg's Arch Pharmacol 304:21–26

Fiszman ML, Stefano FJE (1984) Amphetamine-clonidine interaction on neurotransmission in the vas deferens of the rat. Naunyn-Schmiedeberg's Arch Pharmacol 328:148–153

FitzGerald GA, Watkins J, Dollery CT (1981) Regulation of norepinephrine release by peripheral α_2-receptor stimulation. Clin Pharmacol Ther 29:160–167

Forscher P, Oxford GS, Schulz D (1986) Noradrenaline modulates calcium channels in avian dorsal root ganglion cells through tight receptor-channel coupling. J Physiol (Lond) 379:131–144

Frankhuyzen AL, Mulder AH (1982) Pharmacological characterization of presynaptic α-adrenoceptors modulating [^{3}H]noradrenaline and [^{3}H] 5-hydroxytryptamine release from slices of the hippocampus of the rat. Eur J Pharmacol 81:97–106

Fredholm BB, Lindgren E (1986) Possible involvement of the N_i-protein in the prejunctional inhibitory effect of a stable adenosine analogue (R-PIA) on noradrenaline release in the rat hippocampus. Acta Physiol Scand 126:307–309

French AM, Scott NC (1983) Feedback inhibition of responses of rat vas deferens to twin pulse stimulation. Eur J Pharmacol 86:379–383

Fuder H, Muscholl E, Spemann R (1983) The determination of presynaptic pA_2 values of yohimbine and phentolamine on the perfused rat heart under conditions of negligible autoinhibition. Br J Pharmacol 79:109–119

Fuder H, Bath F, Wiebelt H, Muscholl E (1984) Autoinhibition of noradrenaline release from the rat heart as a function of the biophase concentration. Effects of exogenous α-adrenoceptor agonists, cocaine, and perfusion rate. Naunyn-Schmiedeberg's Arch Pharmacol 325:25–33

Fuder H, Braun HJ, Schimkus R (1986) Presynaptic alpha-2 adrenoceptor activation and coupling of the receptor-presynaptic effector system in the perfused rat heart: affinity and efficacy of phenethylamines and imidazoline derivatives. J Pharmacol Exp Ther 237:237–245

Galloway MP, Westfall TC (1982) The release of endogenous norepinephrine from the coccygeal artery of spontaneously hypertensive and Wistar-Kyoto rats. Circul Res 51:225–232

Galzin AM, Langer SZ (1985) Inhibition by 5,6-dihydroxy-2-dimethylaminotetralin (M7) of noradrenergic neurotransmission in the rabbit hypothalamus: role of alpha-2 adrenoceptors and of dopamine receptors. J Pharmacol Exp Ther 233: 459–465

Galzin AM, Dubocovich ML, Langer SZ (1982) Presynaptic inhibition by dopamine receptor agonists of noradrenergic neurotransmission in the rabbit hypothalamus. J Pharmacol Exp Ther 221:461–471

Galzin AM, Moret C, Langer SZ (1984) Evidence that exogenous but not endogenous norepinephrine activates the presynaptic alpha-2 adrenoceptors on serotonergic nerve endings in the rat hypothalamus. J Pharmacol Exp Ther 228:725–732

Galzin AM, Moret C, Verzier B, Langer SZ (1985) Interaction between tricyclic and nontricyclic 5-hydroxytryptamine uptake inhibitors and the presynaptic 5-hydroxytryptamine inhibitory autoreceptors in the rat hypothalamus. J Pharmacol Exp Ther 235:200–211

Galzin AM, Langer SZ, Pasarelli F (1986) Interaction between 5-HT uptake inhibitors and presynaptic inhibitory 5-HT autoreceptors: comparison of K^+ and electrical depolarization. Br J Pharmacol 87:23P

Gillespie JS (1980) Presynaptic receptors in the autonomic nervous system. In: Szekeres L (ed) Adrenergic activators and inhibitors. Springer, Berlin Heidelberg New York, pp 353–425 (Handbook of experimental pharmacology, vol 54/I)

Göthert M (1977) Effects of presynaptic modulators on Ca^{2+}-induced noradrenaline release from cardiac sympathetic nerves. Naunyn-Schmiedeberg's Arch Pharmacol 300:267–272

Göthert M (1984) Facilitatory effect of adrenocorticotropic hormone and related peptides on Ca^{2+}-dependent noradrenaline release from sympathetic nerves. Neuroscience 11:1001–1009

Göthert M, Hentrich F (1984) Role of cAMP for regulation of impulse-evoked noradrenaline release from the rabbit pulmonary artery and its possible relationship to presynaptic ACTH receptors. Naunyn-Schmiedeberg's Arch Pharmacol 328:127–134

Göthert M, Hentrich F (1986) Further evidence for the involvement of cyclic AMP in Ca^{2+}-dependent, but not Ca^{2+}-independent, noradrenaline release in the rabbit pulmonary artery. Arch int Pharmacodyn Ther 284:85–100

Göthert M, Kollecker P (1986) Subendothelial β_2-adrenoceptors in the rat vena cava: facilitation of noradrenaline release via local stimulation of angiotensin II synthesis. Naunyn-Schmiedeberg's Arch Pharmacol 334:156–165

Göthert M, Pohl IM, Wehking E (1979) Effects of presynaptic modulators on Ca^{2+}-induced noradrenaline release from central noradrenergic neurons. Naunyn-Schmiedeberg's Arch Pharmacol 307:21–27

Göthert M, Schlicker E, Köstermann F (1983) Relationship between transmitter uptake inhibition and effects of α-adrenoceptor agonists on serotonin and noradrenaline release in the rat brain cortex. Naunyn-Schmiedeberg's Arch Pharmacol 322:121–128

Göthert M, Schlicker E, Hentrich F, Rohm N, Zerkowski HR (1984) Modulation of noradrenaline release in human saphenous vein via presynaptic α_2-adrenoceptors. Eur J Pharmacol 102:261–267

Grabowska M, Andén NE (1976) Noradrenaline synthesis and utilization: control by nerve impulse flow under normal conditions and after treatment with alpha-adrenoreceptor blocking agents. Naunyn-Schmiedeberg's Arch Pharmacol 292:53–58

Graham RM, Stephenson WH, Pettinger WA (1980) Pharmacological evidence for a functional role of the prejunctional alpha-adrenoreceptor in noradrenergic neuro-transmission in the conscious rat. Naunyn-Schmiedeberg's Arch Pharmacol 311:129–138

Gripenberg J, Heinonen E, Jansson SE (1980) Uptake of radiocalcium by nerve endings isolated from rat brain: pharmacological studies. Br J Pharmacol 71:273–278

Groß G, Göthert M, Glapa U, Engel G, Schümann HJ (1985) Lesioning of serotoninergic and noradrenergic nerve fibres of the rat brain does not decrease binding of

^{3}H-clonidine and ^{3}H-rauwolscine to cortical membranes. Naunyn-Schmiedeberg's Arch Pharmacol 328:229–235

Grundström N, Andersson RGG, Wikberg JES (1981) Prejunctional alpha$_2$ adrenoceptors inhibit contraction of tracheal smooth muscle by inhibiting cholinergic neurotransmission. Life Sci 28:2981–2986

Guimarães S, Brandão F, Paiva MQ (1978) A study of the adrenoceptor-mediated feedback mechanisms by using adrenaline as a false transmitter. Naunyn-Schmiedeberg's Arch Pharmacol 305:185–188

Hagan RM, Hughes IE (1986) Yohimbine affects the evoked overflow of neurotransmitters from rat brain slices by more than one mechanism. J Pharm Pharmacol 38:195–200

Hamilton CA, Reid JL, Zamboulis C (1982) The role of presynaptic α-adrenoceptors in the regulation of blood pressure in the conscious rabbit. Br J Pharmacol 75:417–424

Hedler L, Starke K, Steppeler A (1983) Release of [^{3}H]-amezinium from cortical noradrenergic axons: a model for the study of the α-autoreceptor hypothesis. Br J Pharmacol 78:645–653

Hedlund H, Andersson KE, Larsson B (1985) Effect of drugs interacting with adrenoreceptors and muscarinic receptors in the epididymal and prostatic parts of the human isolated vas deferens. J Auton Pharmacol 5:261–270

Hedqvist P (1981) Trans-synaptic modulation versus α-autoinhibition of noradrenaline secretion. In: Stjärne L, Hedqvist P, Lagercrantz H, Wennmalm Å (eds) Chemical neurotransmission 75 years. Academic Press, London, pp 223–233

Heepe P, Starke K (1985) α-Adrenoceptor antagonists and the release of noradrenaline in rabbit cerebral cortex slices: support for the α-autoreceptor hypothesis. Br J Pharmacol 84:147–155

Henseling M (1983) The influence of uptake$_2$ on the inhibition by unlabelled noradrenaline of the stimulation-evoked overflow of ^{3}H-noradrenaline in rabbit aorta with regard to surface of amine entry. Naunyn-Schmiedeberg's Arch Pharmacol 324:99–107

Hentrich F, Göthert M, Greschuchna D (1985) Involvement of cAMP in modulation of noradrenaline release in the human pulmonary artery. Naunyn-Schmiedeberg's Arch Pharmacol 330:245–247

Hentrich F, Göthert M, Greschuchna D (1986) Noradrenaline release in the human pulmonary artery is modulated by presynaptic α$_2$-adrenoceptors. J Cardiovasc Pharmacol 8:539–544

Heyndrickx GR, Vilaine JP, Moerman EJ, Leusen I (1984) Role of prejunctional α$_2$-adrenergic receptors in the regulation of myocardial performance during exercise in conscious dogs. Circul Res 54:683–693

Hicks PE, Langer SZ, Macrae AD (1985) Differential blocking actions of idazoxan against the inhibitory effects of 6-fluoronoradrenaline and clonidine in the rat vas deferens. Br J Pharmacol 86:141–150

Hicks PE, Najar M, Vidal M, Langer SZ (1986) Possible involvement of presynaptic α$_1$-adrenoceptors in the effects of idazoxan and prazosin on ^{3}H-noradrenaline release from tail arteries of SHR. Naunyn-Schmiedeberg's Arch Pharmacol 333:354–361

Hieble JP, DeMarinis RM, Fowler PJ, Matthews WD (1986) Selective alpha-2 adrenoceptor blockade by SK&F 86466: in vitro characterization of receptor selectivity. J Pharmacol Exp Ther 236:90–96

Hölting T, Starke K (1986) Receptor protection experiments confirm the identity of presynaptic α$_2$-autoreceptors. Naunyn-Schmiedeberg's Arch Pharmacol 333:262–270

Holz GG, Rane SG, Dunlap K (1986) GTP-binding proteins mediate transmitter inhibition of voltage-dependent calcium channels. Nature 319:670–672

Honda K, Miyata-Osawa A, Takenaka T (1985) α_1-Adrenoceptor subtype mediating contraction of the smooth muscle in the lower urinary tract and prostate of rabbits. Naunyn-Schmiedeberg's Arch Pharmacol 330:16–21

Horn JP, McAfee DA (1980) Alpha-adrenergic inhibition of calcium-dependent potentials in rat sympathetic neurones. J Physiol (Lond) 301:191–204

Hovevei-Sion D, Finberg JPM, Bomzon A, Youdim MBH (1983) Effects of forskolin in rat vas deferens – evidence for facilitatory β-adrenoceptors. Eur J Pharmacol 95:295–299

Idowu OA, Zar MA (1977) The use of rat atria as a simple and sensitive in vitro preparation for detecting pre-synaptic actions of drugs on adrenergic transmission. Br J Pharmacol 61:157P

Ilhan M, Long JP, Cannon JG (1976) The ability of pimozide to prevent inhibition by dopamine analogs of cardioaccelerator nerves in cat hearts. Arch int Pharmacodyn Ther 222:70–80

Illes P (1986) Mechanisms of receptor-mediated modulation of transmitter release in noradrenergic, cholinergic and sensory neurones. Neuroscience 17:909–928

Illes P, Dörge L (1985) Mechanism of α_2-adrenergic inhibition of neuroeffector transmission in the mouse vas deferens. Naunyn-Schmiedeberg's Arch Pharmacol 328:241–247

Illes P, Starke K (1983) An electrophysiological study of presynaptic α-adrenoceptors in the vas deferens of the mouse. Br J Pharmacol 78:365–373

Jackisch R, Werle E, Hertting G (1984) Identification of mechanisms involved in the modulation of release of noradrenaline in the hippocampus of the rabbit in vitro. Neuropharmacology 23:1363–1371

Jakobs KH, Bauer S, Minuth M, Watanabe Y (1985) Inhibitory coupling of α_2-adrenoceptors to adenylate cyclase. In: Lefkowitz RJ, Lindenlaub E (eds) Adrenergic receptors: molecular properties and therapeutic implications. Schattauer, Stuttgart, pp 261–271

Janssens W, Verhaeghe R (1983) Modulation of the concentration of noradrenaline at the neuro-effector junction in human saphenous vein. Br J Pharmacol 79:577–585

Johnston H, Majewski H (1986) Prejunctional β-adrenoceptors in rabbit pulmonary artery and mouse atria: effect of α-adrenoceptor blockade and phosphodiesterase inhibition. Br J Pharmacol 87:553–562

Jonkman FAM, Man PW, Thoolen MJMC, van Zwieten PA (1985) Location of the mechanism of the clonidine withdrawal tachycardia in rats. J Pharm Pharmacol 37:580–582

Kahan T, Hjemdahl P, Dahlöf C (1984) Relationship between the overflow of endogenous and radiolabelled noradrenaline from canine blood perfused gracilis muscle. Acta Physiol Scand 122:571–582

Kalsner S (1979) Single pulse stimulation of guinea-pig vas deferens and the presynaptic receptor hypothesis. Br J Pharmacol 66:343–349

Kalsner S (1981) The role of calcium in the effects of noradrenaline and phenoxybenzamine on adrenergic transmitter release from atria: no support for negative feedback of release. Br J Pharmacol 73:363–371

Kalsner S (1982) Evidence against the unitary hypothesis of agonist and antagonist action at presynaptic adrenoceptors. Br J Pharmacol 77:375–380

Kalsner S (1983) The effects of yohimbine on presynaptic and postsynaptic events during sympathetic nerve activation in cattle iris: a critique of presynaptic receptor theory. Br J Pharmacol 78:247–253

Kalsner S (1985a) Clonidine and presynaptic adrenoceptor theory. Br J Pharmacol 85:143–147

Kalsner S (1985b) Is there feedback regulation of neurotransmitter release by autoreceptors? Biochem Pharmacol 34:4085–4097

Kalsner S, Chan CC (1979) Adrenergic antagonists and the presynaptic receptor hypothesis in vascular tissue. J Pharmacol Exp Ther 211:257–264

Kalsner S, Quillan M (1984) A hypothesis to explain the presynaptic effects of adrenoceptor antagonists. Br J Pharmacol 82:515–522

Katada T, Gilman AG, Watanabe Y, Bauer S, Jakobs KH (1985) Protein kinase C phosphorylates the inhibitory guanine-nucleotide-binding regulatory component and apparently suppresses its function in hormonal inhibition of adenylate cyclase. Eur J Biochem 151:431–437

Kato E, Koketsu K, Kuba K, Kumamoto E (1985) The mechanism of the inhibitory action of adrenaline on transmitter release in bullfrog sympathetic ganglia: independence of cyclic AMP and calcium ions. Br J Pharmacol 84:435–443

Keith RA, Howe BB, Salama AI (1986) Modulation of peripheral beta-1 and alpha-2 receptor sensitivities by the administration of the tricyclic antidepressant, imipramine, alone and in combination with alpha-2 antagonists to rats. J Pharmacol Exp Ther 236:356–363

Kenakin TP (1984) The relative contribution of affinity and efficacy to agonist activity: organ selectivity of noradrenaline and oxymetazoline with reference to the classification of drug receptors. Br J Pharmacol 81:131–141

Kirpekar SM, Furchgott RF, Wakade AR, Prat JC (1973) Inhibition by sympathomimetic amines of the release of norepinephrine evoked by nerve stimulation in the cat spleen. J Pharmacol Exp Ther 187:529–538

Kobinger W (1967) Über den Wirkungsmechanismus einer neuen antihypertensiven Substanz mit Imidazolinstruktur. Naunyn-Schmiedebergs Arch Pharmak exp Path 258:48–58

Kobinger W, Pichler L (1980) Investigation into different types of post- and presynaptic α-adrenoceptors at cardiovascular sites in rats. Eur J Pharmacol 65: 393–402

Kroneberg G, Oberdorf A, Hoffmeister F, Wirth W (1967) Zur Pharmakologie von 2-(2,6-Dimethylphenylamino)-4H-5,6-dihydro-1,3-thiazin (Bayer 1470), eines Hemmstoffes adrenergischer und cholinergischer Neurone. Naunyn-Schmiedebergs Arch Pharmak exp Path 256:257–280

Kubo T, Goshima Y, Ueda H, Misu Y (1986) Diminished α_2-adrenoceptor-mediated modulation of noradrenergic neurotransmission in the posterior hypothalamus of spontaneously hypertensive rats. Neurosci Lett 65:29–34

Laduron PM (1985) Axonal transport of presynaptic receptors. In: Kalsner S (ed) Trends in autonomic pharmacology, vol 3. Taylor and Francis, London, pp 113–127

Lai RT, Watanabe Y, Yoshida H (1983) Effect of islet-activating protein (IAP) on contractile responses of rat vas deferens: evidence for participation of N_i (inhibitory GTP binding regulating protein) in the α_2-adrenoceptor-mediated response. Eur J Pharmacol 90:453–456

Langer SZ (1977) Presynaptic receptors and their role in the regulation of transmitter release. Br J Pharmacol 60:481–497

Langer SZ (1981) Presynaptic regulation of the release of catecholamines. Pharmacol Rev 32:337–362

Langer SZ, Dubocovich ML (1981) Cocaine and amphetamine antagonize the decrease of noradrenergic neurotransmission elicited by oxymetazoline but potentiate the inhibition by α-methylnorepinephrine in the perfused cat spleen. J Pharmacol Exp Ther 216:162–171

Langer SZ, Dubocovich ML, Celuch SM (1975) Prejunctional regulatory mechanisms for noradrenaline release elicited by nerve stimulation. In: Almgren O, Carlsson A, Engel J (eds) Regulation of catecholamine turnover. Chemical tools in catecholamine research, vol II. Elsevier/North-Holland, Amsterdam, pp 183–191

Langer SZ, Adler-Graschinsky E, Giorgi O (1977) Physiological significance of α-adrenoceptor-mediated negative feedback mechanism regulating noradrenaline release during nerve stimulation. Nature 265:648–650

Langley AE, Weiner N (1978) Enhanced exocytotic release of norepinephrine consequent to nerve stimulation by low concentrations of cyclic nucleotides in the presence of phenoxybenzamine. J Pharmacol Exp Ther 205:426−437

Lattimer N, Rhodes KF (1985) A difference in the affinity of some selective α_2-adrenoceptor antagonists when compared on isolated vasa deferentia of rat and rabbit. Naunyn-Schmiedeberg's Arch Pharmacol 329:278−281

Lattimer N, McAdams RP, Rhodes KF, Sharma S, Turner SJ, Waterfall JF (1984) Alpha$_2$-adrenoceptor antagonism and other pharmacological antagonist properties of some substituted benzoquinolizines and yohimbine in vitro. Naunyn-Schmiedeberg's Arch Pharmacol 327:312−318

Ledda F, Mantelli L (1984) Differences between the prejunctional effects of phenylephrine and clonidine in guinea-pig isolated atria. Br J Pharmacol 81:491−497

Leedham JA, Pennefather JN (1982) Dopamine acts at the same receptors as noradrenaline in the rat isolated vas deferens. Br J Pharmacol 77:293−299

Leighton J, Butz KR, Parmeter LL (1979) Effect of α-adrenergic agonists and antagonists on neurotransmission in the rat anococcygeus muscle. Eur J Pharmacol 58:27−38

Levin BE (1984) Axonal transport and presynaptic location of α_2-adrenoreceptors in locus coeruleus neurons. Brain Res 321:180−182

L'Heureux R, Dennis T, Curet O, Scatton B (1986) Measurement of endogenous noradrenaline release in the rat cerebral cortex in vivo by transcortical dialysis: effects of drugs affecting noradrenergic transmission. J Neurochem 46:1794−1801

Limberger N, Starke K (1983) Partial agonist effect of 2-[2-(1,4-benzodioxanyl)]-2-imidazoline (RX 781094) at presynaptic α_2-adrenoceptors in rabbit ear artery. Naunyn-Schmiedeberg's Arch Pharmacol 324:75−78

Limberger N, Starke K (1984) Further study of prerequisites for the enhancement by α-adrenoceptor antagonists of the release of noradrenaline. Naunyn-Schmiedeberg's Arch Pharmacol 325:240−246

Limberger N, Späth L, Hölting T, Starke K (1986) Mutual interaction between presynaptic α_2-adrenoceptors and opioid κ-receptors at the noradrenergic axons of rabbit brain cortex. Naunyn-Schmiedeberg's Arch Pharmacol 334:166−171

Loiacono RE, Rand MJ, Story DF (1985) Interaction between the inhibitory action of acetylcholine and the α-adrenoceptor autoinhibitory feedback system on release of [^{3}H]-noradrenaline from rat atria and rabbit ear artery. Br J Pharmacol 84:697−705

Lokhandwala MF, Steenberg ML (1984) Selective activation by LY-141865 and apomorphine of presynaptic dopamine receptors in the rat kidney and influence of stimulation parameters in the action of dopamine. J Pharmacol Exp Ther 228:161−167

Lorenz RR, Vanhoutte PM, Shepherd JT (1979) Interaction between neuronal amine uptake and prejunctional alpha-adrenergic receptor activation in smooth muscle from canine blood vessels and spleen. Blood Vess 16:113−125

Lundberg JM, Rudehill A, Sollevi A, Theodorsson-Norheim E, Hamberger B (1986) Frequency- and reserpine-dependent chemical coding of sympathetic transmission: differential release of noradrenaline and neuropeptide Y from pig spleen. Neurosci Lett 63:96−100

Madjar H, Docherty JR, Starke K (1980) An examination of pre- and postsynaptic α-adrenoceptors in the autoperfused rabbit hindlimb. J Cardiovasc Pharmacol 2:619−627

Magnan J, Regoli D, Quirion R, Lemaire S, St-Pierre S, Rioux F (1979) Studies on the inhibitory action of somatostatin in the electrically stimulated rat vas deferens. Eur J Pharmacol 55:347−354

Majewski H (1983) Modulation of noradrenaline release through activation of presynaptic β-adrenoreceptors. J Auton Pharmacol 3:47−60

Majewski H, Rand MJ, Tung LH (1981) Activation of prejunctional β-adrenoceptors in rat atria by adrenaline applied exogenously or released as a co-transmitter. Br J Pharmacol 73:669−679

Majewski H, Rump LC, Hedler L, Starke K (1983a) Effects of α_1- and α_2-adrenoceptor blocking drugs on noradrenaline release rate in anesthetized rabbits. J Cardiovasc Pharmacol 5:703−711

Majewski H, Hedler L, Starke K (1983b) Modulation of noradrenaline release in the conscious rabbit through α-adrenoceptors. Eur J Pharmacol 93:255−264

Majewski H, Hedler L, Starke K (1983c) Evidence for a physiological role of presynaptic α-adrenoceptors: modulation of noradrenaline release in the pithed rabbit. Naunyn-Schmiedeberg's Arch Pharmacol 324:256−263

Majewski H, Hedler L, Schurr C, Starke K (1985) Dual effect of adrenaline on noradrenaline release in the pithed rabbit. J Cardiovasc Pharmacol 7:251−257

Malta E, Raper C, Tawa PE (1981) Pre- and postjunctional effects of clonidine- and oxymetazoline-like compounds in guinea-pig ileal preparations. Br J Pharmacol 73:355−362

Markstein R, Digges K, Marshall NR, Starke K (1984) Forskolin and the release of noradrenaline in cerebrocortical slices. Naunyn-Schmiedeberg's Arch Pharmacol 325:17−24

Marshall I (1983) Stimulation-evoked release of [^{3}H]-noradrenaline by 1, 10 or 100 pulses and its modification through presynaptic α_2-adrenoceptors. Br J Pharmacol 78:221−231

Marwaha J, Aghajanian GK (1982) Relative potencies of alpha-1 and alpha-2 antagonists in the locus ceruleus, dorsal raphe and dorsal lateral geniculate nuclei: an electrophysiological study. J Pharmacol Exp Ther 222:287−293

McAfee DA, Henon BK, Horn JP, Yarowsky P (1981) Calcium currents modulated by adrenergic receptors in sympathetic neurons. Fed Proc 40:2246−2249

McCulloch MW, Rand MJ, Story DF (1972) Inhibition of ^{3}H-noradrenaline release from sympathetic nerves of guinea-pig atria by a presynaptic α-adrenoceptor mechanism. Br J Pharmacol 46:523P-524P

McCulloch MW, Rand MJ, Story DF, Sutton I (1981) Prejunctional α-adrenoreceptors subserve a physiological role in cardiac noradrenergic transmission. J Auton Pharmacol 1:407−412

McCulloch MW, Papanicolaou M, Rand MJ (1985) Evidence for autoinhibition of stimulation-induced noradrenaline release from vasa deferentia of the guinea-pig and rat. Br J Pharmacol 86:455−464

Medgett IC, Rand MJ (1981) Dual effects of clonidine on rat prejunctional α-adrenoceptors. Clin Exp Pharmacol Physiol 8:503−507

Medgett IC, McCulloch MW, Rand MJ (1978) Partial agonist action of clonidine on prejunctional and postjunctional α-adrenoceptors. Naunyn-Schmiedeberg's Arch Pharmacol 304:215−221

Meeley MP, Ernsberger PR, Granata AR, Reis DJ (1986) An endogenous clonidine-displacing substance from bovine brain: receptor binding and hypotensive actions in the ventrolateral medulla. Life Sci 38:1119−1126

Minson JB, de la Lande IS (1984) Factors influencing the release of noradrenaline from hypothalamic slices of the possum, Trichosurus vulpecula. Aust J Exp Biol Med Sci 62:341−354

Mishima S, Miyahara H, Suzuki H (1984) Transmitter release modulated by α-adrenoceptor antagonists in the rabbit mesenteric artery: a comparison between noradrenaline outflow and electrical activity. Br J Pharmacol 83:537−547

Mobley P, Greengard P (1985) Evidence for widespread effects of noradrenaline on axon terminals in the rat frontal cortex. Proc Natl Acad Sci USA 82:945−947

Montel H, Starke K, Weber F (1974) Influence of morphine and naloxone on the release of noradrenaline from rat brain cortex slices. Naunyn-Schmiedeberg's Arch Pharmacol 283:357−369

Moore PK, Griffiths RJ (1982) Pre-synaptic and post-synaptic effects of xylazine and naphazoline on the bisected rat vas deferens. Arch int Pharmacodyn Ther 260:70–77

Morita K, North RA (1981) Clonidine activates membrane potassium conductance in myenteric neurones. Br J Pharmacol 74:419–428

Morris MJ, Elghozi JL, Dausse JP, Meyer P (1981) α_1- and α_2-adrenoceptors in rat cerebral cortex: effect of frontal lobotomy. Naunyn-Schmiedeberg's Arch Pharmacol 316:42–44

Mottram DR (1983) Pre-junctional α_2-adrenoceptor activity of B-HT 920. J Pharm Pharmacol 35:652–655

Mulder AH, Schoffelmeer ANM (1985) Catecholamine and opioid receptors, pre-synaptic inhibition of CNS neurotransmitter release, and adenylate cyclase. In: Cooper DMF, Seamon KB (eds) Advances in cyclic nucleotide and protein phosphorylation research, vol 19. Raven, New York, pp 273–286

Mulder AH, Frankhuyzen AL, Stoof JC, Wemer J, Schoffelmeer ANM (1984) Catecholamine receptors, opiate receptors, and presynaptic modulation of neurotransmitter release in the brain. In: Usdin E, Carlsson A, Dahlström A, Engel J (eds) Catecholamines. Part B: Neuropharmacology and central nervous system – theoretical aspects. Liss, New York, pp 47–58

Muramatsu I, Fujiwara M, Ikushima S, Ashida K (1980) Effects of goniopora toxin on guinea-pig blood vessels. Naunyn-Schmiedeberg's Arch Pharmacol 312: 193–197

Murphy MB, Brown MJ, Dollery CT (1984) Evidence for a peripheral component in the sympatholytic actions of clonidine and guanfacine in man. Eur J Clin Pharmacol 27:23–27

Muscholl E (1973) Introduction. In: Proceedings of the 2nd meeting on adrenergic mechanisms. University of Porto, pp 33–39

Nakamura S, Tepper JM, Young SJ, Groves PM (1981) Neurophysiological consequences of presynaptic receptor activation: changes in noradrenergic terminal excitability. Brain Res 226:155–170

Nedergaard OA (1986) Presynaptic α-adrenoceptor control of transmitter release from vascular sympathetic neurones in vitro. In: Grobecker H, Philippu A, Starke K (eds) New aspects of the role of adrenoceptors in the cardiovascular system. Springer, Berlin Heidelberg New York, pp 24–32

Nilsson H, Ljung B, Sjöblom N, Wallin BG (1985a) The influence of the sympathetic impulse pattern on contractile responses of rat mesenteric arteries and veins. Acta Physiol Scand 123:303–309

Nilsson H, Sjöblom N, Folkow B (1985b) Interaction between prejunctional α_2-receptors and neuronal transmitter reuptake in small mesenteric arteries from the rat. Acta Physiol Scand 125:245–252

Nörenberg W (1986) Electrophysiological evidence for an α_2-adrenergic inhibitory control of transmitter release in the rabbit mesenteric artery. Naunyn-Schmiedeberg's Arch Pharmacol 332:R79

North RA (1986) Receptors on individual neurones. Neuroscience 17:899–907

North RA, Surprenant A (1985) Inhibitory synaptic potentials resulting from α_2-adrenoceptor activation in guinea-pig submucous plexus neurones. J Physiol (Lond) 358:17–33

North RA, Williams JT (1984) On the inhibition by opiates of norepinephrine release. In: Usdin E, Carlsson A, Dahlström A, Engel J (eds) Catecholamines. Part B: Neuropharmacology and central nervous system – theoretical aspects. Liss, New York, pp 207–211

Palaty V (1984) Release of 3,4-dihydroxyphenylglycol from the rat tail artery induced by veratridine. Can J Physiol Pharmacol 62:151–152

Pelayo F, Dubocovich ML, Langer SZ (1978) Possible role of cyclic nucleotides in regulation of noradrenaline release from rat pineal through presynaptic adrenoceptors. Nature 274:76–78

Pelayo F, Dubocovich ML, Langer SZ (1980) Inhibition of neuronal uptake reduces the presynaptic effects of clonidine but not of α-methylnoradrenaline on the stimulation-evoked release of ^{3}H-noradrenaline from rat occipital cortex slices. Eur J Pharmacol 64:143—155

Pennefather JN (1983) A study of stimulation-evoked activation of α_2-adrenoceptors in the rat isolated vas deferens. Clin Exp Pharmacol Physiol 10:381—393

Pernow J, Saria A, Lundberg JM (1986) Mechanisms underlying pre- and postjunctional effects of neuropeptide Y in sympathetic vascular control. Acta Physiol Scand 126:239—249

Pizarro M, Valdivieso MP, Orrego F (1986) Differential effects of veratridine and calcium on the release of [^{3}H]noradrenaline and [^{14}C] α-aminoisobutyrate from rat brain cortex slices. Neurochem Int 8:207—212

Polónia JJ, Paiva MQ, Guimarães S (1985) Pharmacological characterization of postsynaptic α-adrenoceptor subtypes in five different dog arteries in-vitro. J Pharm Pharmacol 37:205—208

Powis DA, Baker PF (1986) α_2-Adrenoceptors do not regulate catecholamine secretion by bovine adrenal medullary cells: a study with clonidine. Mol Pharmacol 29:134—141

Rafuse PE, Smith PA (1986) α_2-Adrenergic hyperpolarization is not involved in slow synaptic inhibition in amphibian sympathetic ganglia. Br J Pharmacol 87:409—416

Raiteri M, Maura G, Versace P (1983) Functional evidence for two stereochemically different alpha-2 adrenoceptors regulating central norepinephrine and serotonin release. J Pharmacol Exp Ther 224:679—684

Raiteri M, Bonanno G, Marchi M, Maura G (1984) Is there a functional linkage between neurotransmitter uptake mechanisms and presynaptic receptors? J Pharmacol Exp Ther 231:671—677

Ramme D, Illes P, Späth L, Starke K (1986) Blockade of α_2-adrenoceptors permits the operation of otherwise silent opioid κ-receptors at the sympathetic axons of rabbit jejunal arteries. Naunyn-Schmiedeberg's Arch Pharmacol 334:48—55

Rand MJ, McCulloch MW, Story DF (1980) Catecholamine receptors on nerve terminals. In: Szekeres L (ed) Adrenergic activators and inhibitors. Springer, Berlin Heidelberg New York, pp 223—266 (Handbook of experimental pharmacology, vol 54/I)

Rane SG, Dunlap K (1986) Kinase C activator 1,2-oleoylacetylglycerol attenuates voltage-dependent calcium current in sensory neurons. Proc Natl Acad Sci USA 83:184—188

Reichenbacher D, Reimann W, Starke K (1982) α-Adrenoceptor-mediated inhibition of noradrenaline release in rabbit brain cortex slices. Naunyn-Schmiedeberg's Arch Pharmacol 319:71—77

Remie R, Zaagsma J (1986) A new technique for the study of vascular presynaptic receptors in freely moving rats. Am J Physiol 251:H463—H467

Robie NW (1980) Evaluation of presynaptic α-receptor function in the canine renal vascular bed. Am J Physiol 239:H422—H426

Ruffolo RR (1984) Interactions of agonists with peripheral α-adrenergic receptors. Fed Proc 43:2910—2916

Rump LC, Majewski H (1987) Modulation of noradrenaline release through α_1- and α_2-adrenoceptors in rat isolated kidney. J Cardiovasc Pharmacol (in press)

Ryan LJ, Tepper JM, Sawyer SF, Young SJ, Groves PM (1985) Autoreceptor activation in central monoamine neurons: modulation of neurotransmitter release is not mediated by intermittent axonal conduction. Neuroscience 15:925—931

Saeed M, Holtz J, Elsner D, Bassenge E (1985) Sympathetic control of myocardial oxygen balance in dogs mediated by activation of coronary vascular α_2-adrenoceptors. J Cardiovasc Pharmacol 7:167—173

Saelens DA, Williams PB (1983) Evidence for prejunctional α- and β-adrenoceptors in the canine saphenous vein: influence of frequency of stimulation and external calcium concentration. J Cardiovasc Pharmacol 5:598–603

Sakakibara Y, Fujiwara M, Muramatsu I (1982) Pharmacological characterization of the alpha adrenoceptors of the dog basilar artery. Naunyn-Schmiedeberg's Arch Pharmacol 319:1–7

Savola JM, Virtanen R, Karjalainen A, Ruskoaho H, Puurunen J, Kärki NT (1986) Re-evaluation of drug-interaction with α-adrenoceptors in vivo and in vitro using imidazole derivatives. Life Sci 38:1409–1415

Schlicker E, Göthert M, Köstermann F, Clausing R (1983) Effects of α-adrenoceptor antagonists on the release of serotonin and noradrenaline from rat brain cortex slices. Naunyn-Schmiedeberg's Arch Pharmacol 323:106–113

Schoffelmeer ANM, Mulder AH (1983a) Differential control of Ca^{2+}-dependent [^{3}H] noradrenaline release from rat brain slices through presynaptic opiate receptors and α-adrenoceptors. Eur J Pharmacol 87:449–458

Schoffelmeer ANM, Mulder AH (1983b) [^{3}H] Noradrenaline release from brain slices induced by an increase in the intracellular sodium concentration: role of intracellular calcium stores. J Neurochem 40:615–621

Schoffelmeer ANM, Mulder AH (1983c) ^{3}H-Noradrenaline release from rat neocortical slices in the absence of extracellular Ca^{2+} and its presynaptic alpha$_2$-adrenergic modulation. Naunyn-Schmiedeberg's Arch Pharmacol 323:188–192

Schoffelmeer ANM, Hogenboom F, Mulder AH (1985a) Evidence for a presynaptic adenylate cyclase system facilitating [^{3}H]norepinephrine release from rat brain neocortex slices and synaptosomes. J Neurosci 5:2685–2689

Schoffelmeer ANM, Wardeh G, Mulder AH (1985b) Cyclic AMP facilitates the electrically evoked release of radiolabelled noradrenaline, dopamine and 5-hydroxytryptamine from rat brain slices. Naunyn-Schmiedeberg's Arch Pharmacol 330:74–76

Schoffelmeer ANM, Putters J, Mulder AH (1986a) Activation of presynaptic α$_2$-adrenoceptors attenuates the inhibitory effect of μ-opioid receptor agonists on noradrenaline release from brain slices. Naunyn-Schmiedeberg's Arch Pharmacol 333:377–380

Schoffelmeer ANM, Wierenga EA, Mulder AH (1986b) Role of adenylate cyclase in presynaptic α$_2$-adrenoceptor- and μ-opioid receptor-mediated inhibition of [^{3}H]noradrenaline release from rat brain cortex slices. J Neurochem 46:1711–1717

Sharma TR, Wakade TD, Malhotra RK, Wakade AR (1986) Secretion of catecholamines from the perfused adrenal gland of the rat is not regulated by α-adrenoceptors. Eur J Pharmacol 122:167–172

Shepperson NB, Duval N, Massingham R, Langer SZ (1981) Pre- and postsynaptic alpha adrenoceptor selectivity studies with yohimbine and its two diastereoisomers rauwolscine and corynanthine in the anesthetized dog. J Pharmacol Exp Ther 219:540–546

Skärby T (1984) Pharmacological properties of prejunctional α-adrenoceptors in isolated feline middle cerebral arteries; comparison with the postjunctional α-adrenoceptors. Acta Physiol Scand 122:165–174

Sneddon P, Westfall DP (1984) Pharmacological evidence that adenosine triphosphate and noradrenaline are co-transmitters in the guinea-pig vas deferens. J Physiol (Lond) 347:561–580

Sneddon P, Meldrum LA, Burnstock G (1984) Control of transmitter release in guinea-pig vas deferens by prejunctional P$_1$-purinoceptors. Eur J Pharmacol 105:293–299

Starke K (1972a) Alpha sympathomimetic inhibition of adrenergic and cholinergic transmission in the rabbit heart. Naunyn-Schmiedeberg's Arch Pharmacol 274:18–45

Starke K (1972b) Influence of extracellular noradrenaline on the stimulation-evoked secretion of noradrenaline from sympathetic nerves: evidence for an α-receptor-mediated feed-back inhibition of noradrenaline release. Naunyn-Schmiedeberg's Arch Pharmacol 275:11–23

Starke K (1977) Regulation of noradrenaline release by presynaptic receptor systems. Rev Physiol Biochem Pharmacol 77:1–124

Starke K (1981a) α-Adrenoceptor subclassification. Rev Physiol Biochem Pharmacol 88:199–236

Starke K (1981b) Presynaptic receptors. Ann Rev Pharmacol Toxicol 21:7–30

Starke K, Altmann KP (1973) Inhibition of adrenergic neurotransmission by clonidine: an action on prejunctional α-receptors. Neuropharmacology 12:339–347

Starke K, Langer SZ (1979) A note on terminology for presynaptic receptors. In: Langer SZ, Starke K, Dubocovich ML (eds) Presynaptic receptors. Pergamon, Oxford, pp 1–3

Starke K, Montel H (1974) Influence of drugs with affinity for α-adrenoceptors on noradrenaline release by potassium, tyramine and dimethylphenylpipera-zinium. Eur J Pharmacol 27:273–280

Starke K, Wagner J, Schümann HJ (1972) Adrenergic neuron blockade by clonidine: comparison with guanethidine and local anesthetics. Arch int Pharmacodyn Ther 195:291–308

Starke K, Montel H, Gayk W, Merker R (1974) Comparison of the effects of clonidine on pre-.and postsynaptic adrenoceptors in the rabbit pulmonary artery. Naunyn-Schmiedeberg's Arch Pharmacol 285:133–150

Starke K, Endo T, Taube HD (1975a) Relative pre- and postsynaptic potencies of α-adrenoceptor agonists in the rabbit pulmonary artery. Naunyn-Schmiedeberg's Arch Pharmacol 291:55–78

Starke K, Borowski E, Endo T (1975b) Preferential blockade of presynaptic α-adrenoceptors by yohimbine. Eur J Pharmacol 34:385–388

Steenberg ML, Ekas RD, Lokhandwala MF (1983) Effect of epinephrine on norepine-phrine release from rat kidney during sympathetic nerve stimulation. Eur J Pharmacol 93:137–148

Sternweis PC, Robishaw JD (1984) Isolation of two proteins with high affinity for guanine nucleotides from membranes of bovine brain. J Biol Chem 259: 13806–13813

Stevens MJ, Moulds RFW (1982) Are the pre- and postsynaptic α-adrenoceptors in human vascular smooth muscle atypical? J Cardiovasc Pharmacol 4:S129–S133

Stevens MJ, Moulds RFW (1985) Neuronally released norepinephrine does not preferentially activate postjunctional α_1-adrenoceptors in human blood vessels in vitro. Circul Res 57:399–405

Stjärne L (1973) Michaelis-Menten kinetics of secretion of sympathetic neuro-transmitter as a function of external calcium: effect of graded alpha-adreno-ceptor blockade. Naunyn-Schmiedeberg's Arch Pharmacol 278:323–327

Stjärne L (1978) Facilitation and receptor-mediated regulation of noradrenaline secretion by control of recruitment of varicosities as well as by control of electro-secretory coupling. Neuroscience 3:1147–1155

Stjärne L, Åstrand P (1985) Relative pre- and postjunctional roles of noradrenaline and adenosine 5'-triphosphate as neurotransmitters of the sympathetic nerves of guinea-pig and mouse vas deferens. Neuroscience 14:929–946

Stjärne L, Bartfai T, Alberts P (1979) The influence of 8-Br 3',5'-cyclic nucleotide analogs and of inhibitors of 3',5'-cyclic nucleotide phosphodiesterase, on nor-adrenaline secretion and neuromuscular transmission in guinea-pig vas deferens. Naunyn-Schmiedeberg's Arch Pharmacol 308:99–105

Stjärne L, Alberts P, Bartfai T (1986a) Effects of chloride ion substitution on the frequency dependence and α-autoinhibition of [³H]noradrenaline secretion in guinea-pig vas deferens. Acta Physiol Scand 127:327–333

Stjärne L, Lundberg JM, Åstrand P (1986b) Neuropeptide Y – a cotransmitter with noradrenaline and adenosine 5'-triphosphate in the sympathetic nerves of the mouse vas deferens? A biochemical, physiological and electropharmacological study. Neuroscience 18:151–166

Story DD, Briley MS, Langer SZ (1979) The effects of chemical sympathectomy with 6-hydroxydopamine on α-adrenoceptor and muscarinic cholinoceptor binding in rat heart ventricle. Eur J Pharmacol 57:423–426

Story DF, McCulloch MW, Rand MJ, Standford-Starr CA (1981) Conditions required for the inhibitory feedback loop in noradrenergic transmission. Nature 293:62–65

Story DF, Standford-Starr CA, Rand MJ (1985) Evidence for the involvement of α_1-adrenoceptors in negative feedback regulation of noradrenergic transmitter release in rat atria. Clin Sci 68:111s–115s

Stute N, Trendelenburg U (1984) The outward transport of axoplasmic noradrenaline induced by a rise of the sodium concentration in the adrenergic nerve endings of the rat vas deferens. Naunyn-Schmiedeberg's Arch Pharmacol 327:124–132

Su C, Kubo T (1984) Alpha-adrenoceptor- and prostaglandin-mediated modulation of vascular adrenergic neurotransmission in spontaneously hypertensive rats. Jap J Pharmacol 34:457–463

Sullivan AT, Drew GM (1980) Pharmacological characterisation of pre- and post-synaptic α-adrenoceptors in dog saphenous vein. Naunyn-Schmiedeberg's Arch Pharmacol 314:249–258

Suzuki H (1984) Adrenergic transmission in the dog mesenteric vein and its modulation by α-adrenoceptor antagonists. Br J Pharmacol 81:479–489

Tanaka T, Starke K (1979) Binding of ^{3}H-clonidine to an α-adrenoceptor in membranes of guinea-pig ileum. Naunyn-Schmiedeberg's Arch Pharmacol 309:207–215

Tayo FM (1979) Prejunctional inhibitory α-adrenoceptors and dopaminoceptors of the rat vas deferens and the guinea-pig ileum in vitro. Eur J Pharmacol 58:189–195

Tayo FM, Bevan RD, Bevan JA (1986) Changes in postjunctional α-adrenoceptors during postnatal growth in rabbit arteries. Circul Res 58:867–873

Tepper JM, Groves PM, Young SJ (1985) The neuropharmacology of the auto-inhibition of monoamine release. Trends Pharmacol Sci 6:251–256

Toda N, Okamura T, Nakajima M, Miyazaki M (1984) Modification by yohimbine and prazosin of the mechanical response of isolated dog mesenteric, renal and coronary arteries to transmural stimulation and norepinephrine. Eur J Pharmacol 98:69–78

Török TL, Darvasi A, Salamon Z, Tóth P, Kovács A, Nguyen TT, Magyar K (1985) Presynaptic autoinhibition during rest and sodium-pump inhibition in isolated rat portal vein preparation. Neuroscience 16:439–449

Tsukahara T, Taniguchi T, Usui H, Miwa S, Shimohama S, Fujiwara M, Handa H (1986) Sympathetic denervation and alpha adrenoceptors in dog cerebral arteries. Naunyn-Schmiedeberg's Arch Pharmacol 334:436–443

Uchida W, Kimura T, Satoh S (1984) Presence of presynaptic inhibitory α_1-adrenoceptors in the cardiac sympathetic nerves of the dog: effects of prazosin and yohimbine on sympathetic neurotransmission to the heart. Eur J Pharmacol 103:51–56

Ueda H, Goshima Y, Misu Y (1983) Presynaptic mediation by α_2-, β_1- and β_2-adrenoceptors of endogenous noradrenaline and dopamine release from slices of rat hypothalamus. Life Sci 33:371–376

U'Prichard DC (1984) Biochemical characteristics and regulation of brain α_2-adrenoceptors. Ann NY Acad Sci 430:55–75

Vizi ES (1979) Presynaptic modulation of neurochemical transmission. Progr Neurobiol 12:181–290

Vizi ES, Somogyi GT, Hadházy P, Knoll J (1973) Effect of duration and frequency of stimulation on the presynaptic inhibition by α-adrenoceptor stimulation of the adrenergic transmission. Naunyn-Schmiedeberg's Arch Pharmacol 280:79–91

Vizi ES, Ludvig N, Rónai AZ, Folly G (1983) Dissociation of presynaptic α_2-adreno-
    ceptors following prazosin administration: presynaptic effect of prazosin.
    Eur J Pharmacol 95:287—290

Vizi ES, Harsing LG, Zimanvi I, Gaal G (1985) Release and turnover of noradrenaline
    in isolated median eminence: lack of negative feedback modulation. Neuro-
    sience 16:907—916

von Euler US (1979) General views on the relevance of presynaptic receptor systems.
    In: Langer SZ, Starke K, Dubocovich ML (eds) Presynaptic receptors. Pergamon,
    Oxford, pp 5—9

von Kügelgen I, Starke K (1985) Noradrenaline and adenosine triphosphate as
    co-transmitters of neurogenic vasoconstriction in rabbit mesenteric artery.
    J Physiol (Lond) 367:435—455

Wakade AR, Wakade TD (1981) Release of noradrenaline by one pulse: modulation
    of such release by alpha-adrenoceptor antagonists and uptake blockers. Naunyn-
    Schmiedeberg's Arch Pharmacol 317:302—309

Wakade AR, Wakade TD (1983) Mechanism of negative feed-back inhibition of
    norepinephrine release by alpha-adrenergic agonists. Neuroscience 9:673—677

Wakade AR, Wakade TD (1984) Effects of desipramine, trifluoperazine and other
    inhibitors of calmodulin on the secretion of catecholamines from the adrenal
    medulla and postganglionic sympathetic nerves of the salivary gland. Naunyn-
    Schmiedeberg's Arch Pharmacol 325:320—327

Wakade AR, Malhotra RK, Wakade TD (1985) Phorbol ester, an activator of protein
    kinase C, enhances calcium-dependent release of sympathetic neurotransmitter.
    Naunyn-Schmiedeberg's Arch Pharmacol 331:122—124

Warming SE, Shipley SD, Leedham JA, Hartley ML, Handberg GM, Pennefather JN
    (1982) The influence of neuronal uptake upon the relative potencies of agonists
    acting at prejunctional α_2-adrenoceptors in the rat isolated vas deferens. Arch
    int Pharmadocyn Ther 259:14—30

Warnock P, Hyland L, Docherty JR (1985) Further examination of the inhibitory
    actions of α_1-adrenoceptor agonists in rat vas deferens. Eur J Pharmacol 113:
    239—245

Wemer J, van der Lugt JC, de Langen CDJ, Mulder AH (1979) On the capacity of
    presynaptic alpha receptors to modulate norepinephrine release from slices
    of rat neocortex and the affinity of some agonists and antagonists for these
    receptors. J Pharmacol Exp Ther 211:445—451

Wemer J, Schoffelmeer ANM, Mulder AH (1981) Studies on the role of Na^+, K^+ and
    Cl^- ion permeabilities in K^+-induced release of ^3H-noradrenaline from rat brain
    slices and synaptosomes and in its presynaptic α-adrenergic modulation. Naunyn-
    Schmiedeberg's Arch Pharmacol 317:103—109

Wemer J, Schoffelmeer ANM, Mulder AH (1982) Effects of cyclic AMP analogues
    and phosphodiesterase inhibitors on K^+-induced $[^3H]$noradrenaline release from
    rat brain slices and on its presynaptic α-adrenergic modulation. J Neurochem
    39:349—356

Westfall TC (1977) Local regulation of adrenergic neurotransmission. Physiol Rev
    57:659—728

Westfall TC, Xue CY, Carpentier S, Meldrum MJ (1985) Modulation of noradrenaline
    release by presynaptic adrenoceptors in experimental hypertension. In: Szabadi E,
    Bradshaw CM, Nahorski SR (eds) Pharmacology of adrenoceptors. VCH, Wein-
    heim, pp 177—186

Wetzel GT, Goldstein D, Brown JH (1985) Acetylcholine release from rat atria
    can be regulated through an α_1-adrenergic receptor. Circul Res 56:763—766

Wikberg JES (1979) The pharmacological classification of adrenergic α_1 and α_2
    receptors and their mechanisms of action. Acta Physiol Scand, Suppl 468

Williams JT, North RA (1985) Catecholamine inhibition of calcium action potentials
    in rat locus coeruleus neurones. Neuroscience 14:103—109

Williams JT, Henderson G, North RA (1985) Characterization of α_2-adrenoceptors which increase potassium conductance in rat locus coeruleus neurones. Neuroscience 14:95–101

Yamaguchi I, Kopin IJ (1980) Differential inhibition of alpha-1 and alpha-2 adrenoceptor-mediated pressor responses in pithed rats. J Pharmacol Exp Ther 214:275–281

Yamaguchi N (1982) Evidence supporting the existence of presynaptic α-adrenoceptors in the regulation of endogenous noradrenaline release upon hepatic sympathetic nerve stimulation in the dog liver in vivo. Naunyn-Schmiedeberg's Arch Pharmacol 321:177–184

Yamaguchi N, de Champlain J, Nadeau RA (1977) Regulation of norepinephrine release from cardiac sympathetic fibers in the dog by presynaptic α- and β-receptors. Circul Res 41:108–117

Yorikane R, Kanda A, Kimura T, Satoh S (1986) Effects of epinephrine, isoproterenol and IPS-339 on sympathetic transmission to the dog heart: evidence against the facilitatory role of presynaptic beta adrenoceptors. J Pharmacol Exp Ther 238:334–340

Zimmerman BG, Liao JC, Gisslen J (1971) Effect of phenoxybenzamine and combined administration of iproniazid and tropolone on catecholamine release elicited by renal sympathetic nerve stimulation. J Pharmacol Exp Ther 176:603–610

Zukowska-Grojec Z, Bayorh MA, Kopin IJ (1983) Effect of desipramine on the effects of α-adrenoceptor inhibitors on pressor responses and release of norepinephrine into plasma of pithed rats. J Cardiovasc Pharmacol 5:297–301

Note Added in Proof

Saphenous veins and small jejunal arteries of rabbits are further tissues in which noradrenaline release is modified by α_2-adrenoceptor agonists and antagonists (Levitt and Hieble, Eur J Pharmacol 132, 197–205, 1986; Ramme et al., see References). The existence of α_2-autoreceptors in rabbit aorta has been confirmed by experiments in which the electrically evoked release of ^{3}H-adrenaline was measured (Abrahamsen and Nedergaard, Acta pharmacol toxicol 59, 416–424, 1986). The binding of ^{3}H-p-aminoclonidine to brain membranes has been investigated in search for an imidazoline binding site insensitive to phenylethylamines; in fact, in membranes from rat ventrolateral medulla 30% of the total binding of ^{3}H-p-aminoclonidine was not inhibited by phenylethylamines but was displaced by compounds possessing an imidazole ring (Ernsberger, Meeley, Mann and Reis, Eur J Pharmacol 134, 1–13, 1987). In addition to the receptor interactions mentioned in Sect. 7, α_2-autoreceptors also seem to interact with presynaptic adenosine receptors, possibly through competition for one pool of G proteins (Allgaier, Hertting and von Kügelgen, Br J Pharmacol 90, 403–412, 1987). The view, however, that an inhibition of adenylate cyclase might be the next step has again been questioned (Johnston and Majewski, Br J Pharmacol 89, 790P, 1986). The properties of synapsin I have recently been reviewed, including its possible role in presynaptic receptor transduction mechanisms (De Camilli and Greengard, Biochem Pharmacol 35, 4349–4357, 1986). Postsynaptic response studies have supported further the autoreceptor function of presynaptic α_2-receptors in isolated tissues (rat vas deferens: García-Sevilla and Zubieta, Br J Pharmacol 89, 673–683, 1986; rat tail artery: Papanicolaou and Medgett, Eur J Pharmacol 131, 211–218, 1986); the study in vas deferens also shows desensitization of the autoreceptors upon chronic administration of antidepressant drugs. Soma-dendritic α_2-receptors of postganglionic sympathetic neurons, in contrast to presynaptic α_2-receptors, may not be physiological targets of endogenous noradrenaline (Andén, Grabowska-Andén and Nilsson, Naunyn-Schmiedeberg's Arch Pharmacol 335, 40–43, 1987). The operation of presynaptic α_2-autoreceptors in the in situ blood-perfused dog gracilis muscle has been confirmed; these neurons may possess presynaptic α_1-adrenoceptors as well (Kahan and Hjemdahl, Eur J Pharmacol 133, 9–20, 1987). Clonidine and yohimbine caused marked changes in the electrically evoked release of neuropeptide Y from the sympathetic nerves of pithed guinea pigs, indicating α_2-autoreceptor modulation of the release of noradrenaline as well as of this cotransmitter (Dahlöf, Dahlöf and Lundberg, Eur J Pharmacol 131, 279–283, 1986).

Rev. Physiol. Biochem. Pharmacol., Vol. 107
© by Springer-Verlag 1987

Damage to Mammalian Cells by Proteins that Form Transmembrane Pores

SUCHARIT BHAKDI[1] and JØRGEN TRANUM-JENSEN[2]

Contents

[1] Institute of Medical Microbiology, University of Giessen, D-6300 Giessen, Federal Republic of Germany
[2] Anatomy Department C, University of Copenhagen, DK-2200 Copenhagen N, Denmark

1 Introductory Overview

Mammalian cell membranes consist of single bilayers of lipid with integral membrane proteins inserted into the bilayer and more loosely associated peripheral membrane proteins. Such membranes are only 8–9 nm thick, yet they hold together cells whose diameters may reach 7500–12 000 nm. Macromolecules in the cytosol exert a colloid osmotic pressure greater than that exerted by the proteins in the extracellular fluid. As water freely permeates lipid bilayers, cell membranes must balance out the surplus of internal colloid osmotic effectors by actively pumping ions out of the cell. As a net result of active and passive ion movements the cells acquire a characteristic "milieu interieur", the cytoplasm being relatively rich in K^+ and low in Na^+ (and Ca^{2+}). Maintenance of the intracellular ionic composition is required for many cellular functions, and is possible because the membrane-bound ion pumps are able to cope with the slow passive ion fluxes that occur across the intact bilayer. Any perturbation of the permeability barrier which allows free transmembrane diffusion of ions threatens the very existence of a cell. A lesion that cannot be repaired or removed from the membrane will be lethal if it is of a critical size that allows rapid, passive transmembrane ion flux; the critical size will depend on the capacity of the cell to balance the leak by compensatory active pumping of ions. This situation is encountered in cells such as erythrocytes that lack mechanisms for endocytic uptake of membrane areas. In contrast, nucleated cells are generally able to segregate a limited number of permeable sites from their surface and so a certain number of membrane lesions may be generated without causing cell death. However, sublethal damage may trigger pathophysiological events with profound effects on the attacked cells and their environment.

There are several ways for biological molecules to create permeable sites in cell membranes. Enzymatic attack on membrane proteins is invariably restricted to the hydrophilic, extramembranous domains and hence does not primarily perturb the integrity of the lipid bilayer. Enzymatic attack on membrane lipids may occasionally result in permeability defects if cleavage of the target molecules alters their physical properties such that they are no longer able to maintain their orientation in the bilayer. Phospholipase C, which is produced by a number of bacteria and is also an important constituent of several snake venoms, is an example of such an enzyme (for a review of this topic; see Möllby 1978).

Another possible cause of perturbation of membrane integrity involves the physical derangement of membrane lipids (and integral membrane proteins) by the action of surface-active substances (detergents) that spontaneously intercalate into lipid bilayers. Small permeability defects occur when sufficient detergent molecules enter the bilayer; above a critical concentration mixed detergent-lipid micelles form, leading to disintegration of bilayer structure (Helenius and Simons 1975; Tanford and Reynolds 1976). Lysolecithin is the major naturally occurring lipid detergent, but there is currently no evidence that its generation causes any overt membrane damage. Amphiphilic structures and detergent-like properties have been observed in a number of membrane-damaging polypeptides such as mellitin (Haberman 1972; Bernheimer and Rudy 1986) and staphylococcal δ-toxin (McCartney and Arbuthnott 1978; Möllby 1983), and the possibility that their action on cells derives from a detergent-like effect is being considered. The polyene antibiotics that are produced by various species of streptomyces, e.g., the amphotericins filipin and nystatin, are a special group of amphiphilic membrane perturbers (Kinsky 1970; Norman et al. 1976; Behnke et al. 1986a, b). While detergents are intercalated among membrane lipids in a largely random fashion, these polyene antibiotics form discrete, ordered aggregates that create pores of specific size.

A final possible cause of primary perturbation of membrane integrity is the formation of stable transmembrane pores in the bilayer. The concept of pore formation by proteinaceous substances originated from the work on gramicidin A (Urry 1971; Hladky and Haydon 1972; Myers and Haydon 1972; for a recent review see Bernheimer and Rudy 1986). This microbial pentadecapeptide is produced by various strains of *Bacillus brevis* and appears to form dimeric channels in membrane targets which have been studied extensively. Another group of peptides, the alamethicins produced by the fungus *Trichoderma viride,* act by forming voltage-dependent, cation-selective channels through aggregation in the bilayer (Latorre and Alvarez 1981).

The first concrete formulation of the concept that proteins might also damage target membranes in a manner analogous to gramicidin A was made by Mayer in 1972 to account for the mechanism of cytolysis by complement (Mayer 1972). Mayer correctly surmised that such transmembrane pore formation could only be accomplished by the insertion of protein chains into the bilayer. The inserted proteins would have to possess a hydrophobic surface interacting with lipid (as established for integral membrane proteins), and a hydrophilic surface that would permit transmembrane passage of ions and water. In the original "doughnut" hypothesis, it was proposed that terminal complement components formed circular structures identical to the ultrastructural complement lesions, leakage occurring through the protein-lined pores. Today, it is increasingly being recognized that although many protein pores display this basic feature, some do not require a complete circular protein "wall", and may instead be partially lined by an edge of free lipid (Bhakdi and Tranum-Jensen 1984b, 1986).

The important novel proposal of the protein pore concept was that hydrophilic proteins could transform themselves into amphiphilic moieties and become firmly embedded within the apolar regions of lipid bilayers (Mayer 1972). In 1972 the fluid-mosaic model had just been established (Singer and Nicholson 1972), and it was generally believed that proteins were produced either as hydrophilic entities (i.e., secreted proteins) or as amphiphilic polypeptides that remained anchored in the bilayer (i.e., integral membrane proteins). That hydrophilic proteins might undergo a transition to an amphiphilic state to spontaneously insert themselves into a membrane was therefore an audacious proposal that was initially met with skepticism (e.g., Müller-Eberhard 1975). Nevertheless, Mayer's deductive reasoning led him to advance this concept since it alone could account for all the experimental data previously obtained on the complement lesion. In particular, because of the semifluid state of lipids, membrane-inserted proteins with the correct conformation would be the only molecules theoretically capable of generating stable pores across the bilayer.

Verification of the pore concept of complement action involved sophisticated functional analyses with complement proteins on the one hand, and combined biochemical and ultrastructural characterization of the pores after their isolation from target membranes on the other. Awareness of the pore mechanism quickly led to the recognition that it operated in other biological systems, and investigations were accordingly carried out on bacterial toxins and in the field of lymphocyte-mediated cytotoxicity. The objective of this review is to discuss data that have accumulated in this rapidly growing area. The first sections will deal with approaches to the functional, structural, and biochemical characterization of protein pores. Thereafter, the salient features of a number of pores that have been studied to date will be summarized. Efforts will be made to illuminate the biological significance of the interactions of these proteins with the target cells.

2 General Features of Pore Formation

2.1 Types of Protein Pores

Pore sizes can differ widely. To date, the smallest protein pores forming in mammalian target membranes have an effective diameter of about

1–2 nm. Examples of such pores are staphylococcal α-toxin (Füssle et al. 1981; Bhakdi et al. 1984a), *Escherichia coli* hemolysin (Bhakdi et al. 1986), and the cytolysin of *Pseudomonas aeruginosa* (Lutz et al. 1972). Complement pores are generally larger (5–7 nm; Michaels et al. 1976; Giavedoni et al. 1979; Ramm et al. 1982b; 1985). The largest presently recognized pores are formed by the sulfhydryl-activated cytolysins (e.g., streptolysin-O); their diameters may approach 30–35 nm (Buckingham and Duncan 1983; Bhakdi et al. 1985).

Pores may display a characteristic ultrastructure, or they may be undetectable by current electron-microscopic techniques. In the former case, pores are seen as partially or fully circular structures partially buried within the lipid bilayer. They represent stable, macromolecular oligomers of the native, monomeric components, and they can be isolated as such after solubilization of membranes with mild, nondenaturing detergents. Membrane constituents have never been found to be incorporated into the oligomeric protein structures. The compositions of pores devoid of recognizable structure have generally been less well characterized. It now appears that some pores may form solely through insertion of protein monomers into the bilayer (e.g., *E. coli* hemolysin; Bhakdi et al. 1986). Other pores without visible ultrastructure may arise through metastable dimerization of membrane-inserted polypeptides, possibly giving rise to fluctuating pores, as appears to be the case for gramicidin A.

Pores may be homogenous or heterogeneous. Homogeneous, structured pores are · formed when a constant number of protomers constitutes each lesion, as is probably the case for the majority of staphylococcal α-toxin pores, which are hexamers of the molecule (Bhakdi et al. 1981). Homogeneity would also be expected with unstructured pores when lesions are induced by toxin monomers. Structured, oligomeric pores are more often heterogeneous due to the presence of varying numbers of protomers that form the individual lesions. Typical examples are the lesions formed by sulfhydryl-activated cytolysins (streptolysin-O) and complement (Bhakdi and Tranum-Jensen 1986). These lesions are characteristically seen in the electron microscope as partially or fully circular structures with varying diameters, and functional heterogeneity of pore size can be detected by appropriate assays.

2.2 Mode of Pore Formation

The events that induce binding, oligomerization, and insertion in the membrane of pore-forming proteins, as well as those responsible for cessation of the processes, are not clearly understood. There are proteins which appear to bind to specific membrane binder molecules (e.g., cholesterol in the case of sulfhydryl-activated toxins; Alouf 1980; Smyth and Duncan 1978), but specific binding substrates have not been identified for many pore formers, including the complement C5b–9 complex and staphylococcal α-toxin. Oligomerization can be envisaged as following one of two modes. In the first, protein monomers bind to the membrane; lateral movement of these molecules in the membrane then leads to their collision and this in turn might trigger the conformational changes responsible for exposure of hydrophobic domains (e.g., through unfolding of the molecules). This type of oligomeric pore formation has been proposed for streptolysin-O (Hugo et al. 1986) and for staphylococcal α-toxin (Harshman and Sugg 1985). A second mechanism for initiation of the oligomerization process may involve primary attachment of a first molecule to a membrane site, with the concomitant exposure of a novel binding site for attachment of the next protomer to the initially bound protein. This mechanism probably operates in the case of the complement lesion (Podack et al. 1982; Silversmith and Nelsestuen 1986; Bhakdi and Tranum-Jensen 1986).

The insertion of oligomerizing protein pores in membranes appears to occur spontaneously after exposure of apolar regions on the protein complexes, because of the energetically favored hydrophobic inter-actions of these areas with membrane lipids. No specific requirements such as membrane potential have been identified (Henkart and Blumen-thal 1975; Michaels et al. 1976; Menestrina 1986). The insertion of non-oligomerizing pore formers in membranes may follow a different pattern. It has recently been found that *E. coli* hemolysin will only form pores in planar lipid membranes in the presence of a transmembrane electrical potential (G. Menestrina, unpublished work). In this respect, the hemolysin displays highly interesting similarities to pore-forming colicins (Schein et al. 1978; Konisky 1982; Cramer et al. 1983) and to alamethicin (Latorre and Alvarez 1981).

The fate of membrane constituents during the process of pore formation has not been clarified in detail. One possibility is that pore-forming proteins may expel lipid (and integral membrane proteins) at their sites of insertion in the membrane. Some evidence for this mode of pore formation has been found for complement (Kinoshita et al. 1977; Inoue et al. 1977; Shin et al. 1978). Alternatively, the hydro-philic surfaces of the inserted polypeptide might laterally repel lipids and membrane proteins. In this case, expulsion of membrane con-stituents into the extramembranous environment would not be required, but insertion of a sufficient number of pores would be expected to cause nonosmotic membrane expansion and increase membrane surface area; this has been observed following formation of complement lesions in erythrocyte membranes (Valet and Opferkuch 1985). Furthermore, we have been unable to detect the expulsion of significant amounts of membrane lipid into the fluid phase despite extensive pore formation in erythrocyte membranes by streptolysin-O, staphylococcal α-toxin, and complement (unpublished data). Therefore, we currently favor the second possible mode of pore formation over the first.

2.3 Membrane Orientation and Stability of Pores

Membrane-bound protein pores have always been found to expose part of their surface at the external membrane surface. We suspect that partial penetration of a relatively large water-filled pore, inserted for example into the external membrane monolayer, would probably suffice to induce a transmembrane leak. Hence, pore-forming proteins probably need not span the entire thickness of the bilayer. The most detailed analyses of the depth of bilayer penetration have been made on the complement pore, where it appears that the C9 complement component spans the entire bilayer and an epitope appears to have been detected at the cytoplasmic membrane surface (Morgan et al. 1984; however, see discussion on p. 188).

   A general feature of possible significance is the extreme stability of oligomeric membrane-bound pores. All pores studied to date remain antigenically, structurally, and functionally intact despite extensive proteolytic attack at neutral pH. Hence, oligomeric pore formers in their membrane-bound (but not native) state are remarkably stable molecules whose elimination may require the total removal of afflicted membrane areas (e.g., by endocytosis or phagocytosis).

2.4 Molecular Basis of Pore Formation

The primary structures of three pore formers, namely staphylococcal α-toxin, E. coli hemolysin, and the C9 complement component, have been elucidated at a molecular genetical level (Gray and Kehoe 1984; DiScipio et al. 1984; Felmlee et al. 1985; Stanley et al. 1985), and the primary structures of others will probably be determined in the near future. Nevertheless, a molecular model for the formation of the intramembranous domains has not been proposed for any pore former. The surface of the pore that interacts with lipid must be hydrophobic; the surface that permit passage of ions and water must be hydrophilic. Therefore, a stretch of apolar amino acids as found for integral membrane proteins is not to be expected, and the intramembranous domains need not exhibit α-helical structure. Elucidation of the three-dimensional structure of pores will clearly require extensive biophysical analyses in the future.

3 Isolation and Characterization of Pore-Formers: General Comments

3.1 Native Proteins

Native proteins can generally be isolated using conventional procedures for protein fractionation. Purification requires that suitable methods are available for the identification of proteins. Functional assays are usually based on hemolysis measurements, and are particularly recommended when pore formation is elicited by a single molecular species (e.g., with bacterial toxins). The assays are more difficult to perform when the pores are complex in composition and consist of multiple subunits, e.g., in the case of complement components. When specific antisera are available, immunological analyses are often excellent alternative means for monitoring protein fractionations. To obtain such specific antisera, it is often recommended that pores be first isolated from target membranes, and then used as antigens.

   Purification of the native proteins often poses problems due to loss of hemolytic activity, either because of proteolytic degradation or because of spontaneous inactivation. Counteractive measures include the addition of appropriate protease inhibitors in the buffers, the use of fractionation procedures that lead to early removal of protease contaminants, and fractionation at a low temperature. Sometimes, loss of functional activity is not due to proteolysis, but to an alteration of the physical state of the protein (e.g., *E. coli* hemolysin). It is noteworthy that most pore formers tend to aggregate spontaneously, sometimes thereby forming the same structures that are found on target membranes. This property, which has been described for staphylococcal α-toxin (Arbuthnott et al. 1967), C9 (Podack and Tschopp 1982a, b; Tschopp 1984), tetanolysin (Rottem et al. 1982), and perfringolysin (Mitsui et al. 1979a, b), is probably due to the presence of hydrophobic domains that tend to become exposed when the molecules gain contact with each other. Divalent metal ions such as Zn^{2+} may greatly accelerate such spontaneous interactions (Tschopp, 1984), so inclusion of EDTA in isolation buffers may be generally recommendable.

3.2 Membrane-Bound-Proteins

3.2.1 Isolation

It is often much easier to isolate the pore-forming proteins from attacked target membranes for two reasons. First, the membranes will generally absorb the proteins selectively and the bulk of primary contaminants may be removed by washing the lysed membranes. There remains the problem of separating the pore proteins from native membrane constituents. This is easily achieved if the pore formers are of the oligomerizing type and if erythrocytes can be used as lysis targets. The majority of membrane proteins are present in relatively low molecular weight form (relative molecular mass, $M_r < 5 \times 10^5$) in detergent solution, and may thus be separated from macromolecular pores ($M_r \gtrsim 10^6$) by simple procedures such as sucrose density gradient centrifugation or gel chromatography. Membrane solubilization is achieved with appropriate detergents; high concentrations (125 mM) of deoxycholate (DOC) serve

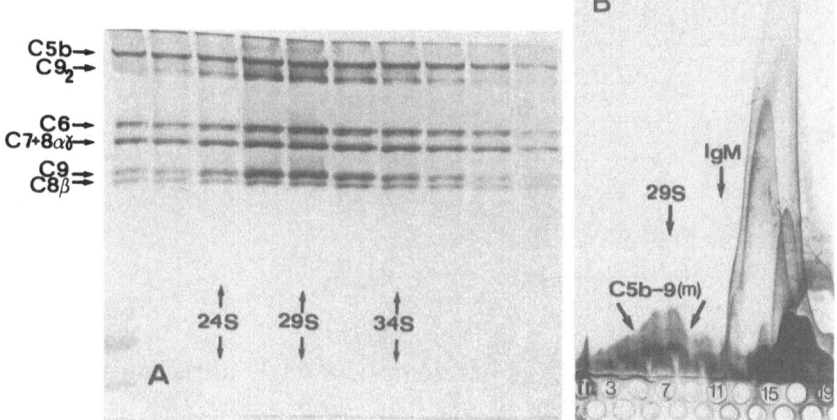

Fig. 1A, B. Rabbit erythrocytes were lysed with whole human serum and the washed membranes solubilized in 250 mM DOC. The solubilisate was centrifuged overnight through a linear sucrose density gradient containing 6.25 mM detergent, and the fractions obtained were analyzed by fused rocket immunoelectrophoresis developed with a polyspecific antiserum to human serum proteins (**B**) and by SDS-PAGE A fractions 2–11 of gradient only). Sedimentation was from *right* to *left;* fraction numbers are given and positions corresponding to 24S, 29S and 34S depicted. C5b–9 terminal membrane complex representing the complement pores sediment as a broad protein peak covering 22S–40S. Membrane and membrane-bound proteins other than C5b–9 are separated by the density gradient ultracentrifugation and are not significant contaminants. The designations of the C5–C9 components are based on published work (Kolb and Müller-Eberhard 1975a; Bhakdi et al. 1976; Ware and Kolb 1981) (From Bhakdi and Tranum-Jensen 1984a)

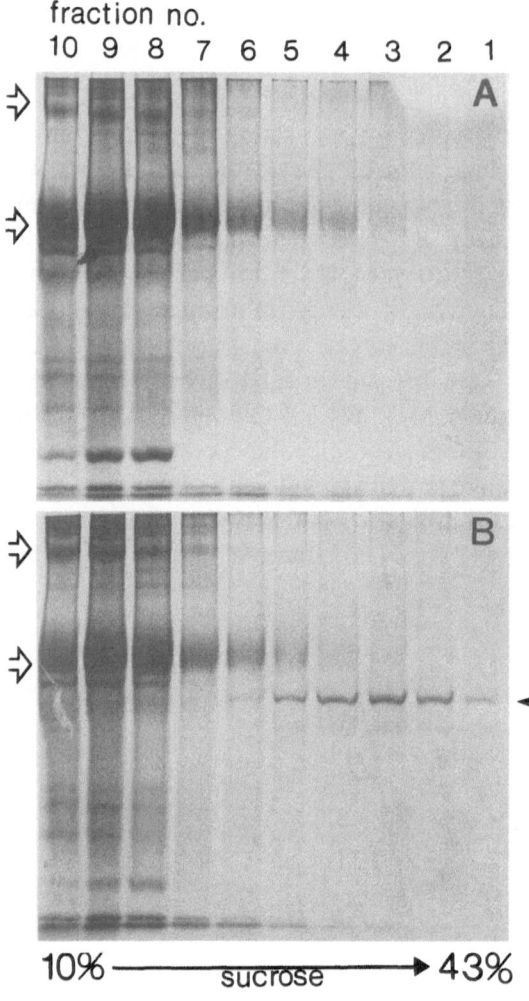

fraction no.
10 9 8 7 6 5 4 3 2 1

Fig. 2A, B. Isolation of streptolysin-O oligomers from rabbit erythrocyte membranes. Control or toxin-treated membranes were solubilized in 250 mM DOC and centrifuged through linear, detergent-containing sucrose density gradients (direction of sedimentation *left* to *right*). Ten equal fractions were collected and samples were analyzed by SDS-PAGE. **A** Control membranes; **B** Streptolysin-O-treated membranes. The streptolysin-O bands (*solid arrows* on the *right*) are observed in the high molecular weight regions (fractions *1* and *6*) separated from the bulk of contaminating erythrocyte membrane proteins. *Open arrows* point to the spectrins and band 3 protein (From Bhakdi et al. 1985)

10% ──────── sucrose ────→ 43%

this purpose excellently. For example, a single ultracentrifugation step in a linear sucrose density gradients with a low DOC concentration (6–10 mM) leads to the isolation of pores formed by complement and sulfhydryl-activated cytolysins from DOC solubilizates (Figs. 1, 2). The initial preparation of lysed membranes does not even require the use of purified complement or toxin components; erythrocytes can simply be treated with whole serum (for isolation of the complement pores) or with partially purified toxin preparations. Since the protein pores are generally resistant to tryptic attack, preparation of very pure protein can sometimes be achieved by proteolysis of the erythrocyte membranes before membrane solubilization. The proteins recovered from a sucrose density gradient are present in extensively delipidated form, the lipid being recovered floating in the detergent micelles at the top of the gradients.

3.2.2 Immunological Characterization

Purified pore formers can be used directly as antigens for raising poly- or monoclonal antibodies. The antisera will react with common epitopes of the native proteins, and will sometimes also exhibit specificities against novel epitopes that are characteristic of the assembled pore structures. Such "neoantigens" have been well characterized for the complement pore. They represent useful immunological markers for differentiation between native proteins and their assembled pore counterparts.

3.2.3 Physicochemical Properties

The native proteins are water-soluble molecules and their molecular weights can generally be determined by conventional procedures. Ideally, molecular weights should be determined by sedimentation equilibrium ultracentrifugation. This is possible if native proteins can be obtained in a nonaggregated state at appropriate concentrations. Reliable molecular-weight determinations have been obtained for the individual components C5–C9 (Müller-Eberhard 1975; Biesecker and Müller-Eberhard 1980; Steckel et al. 1980), staphylococcal α-toxin (Bhakdi et al. 1981), and streptolysin-O (Bhakdi et al. 1984b). Similar analyses have not been possible with *E. coli* hemolysin because of aggregation artifacts. Direct determinations of the polypeptide portions of the proteins can, of course, be obtained when complete complementary DNA (cDNA) sequences become available (e.g., *E. coli* hemolysin; Felmlee et al. 1985).

The determination of the molecular weight of the pore structures is much more difficult because most oligomeric pores are heterogeneous and also tend to aggregate, even in detergent solution. Reliable determinations have only been obtained for staphylococcal α-toxin pores (M_r 2×10^5) and for completely circular C9 oligomers ($M_r \approx 10^6$). The former determinations used ultracentrifugation and correction for bound detergent (Bhakdi et al. 1981), and were recently confirmed by cross-linking studies (Tobkes et al. 1985). The C9 determinations were by analysis of electron scatter of individual molecules (Tschopp et al. 1984). Recently, light-scattering intensity has been used to determine the molecular weight of C5b-9 complexes forming on liposomal membranes (Silversmith and Nelsestuen 1986).

Another way of estimating molecular weights is by calculating the protein mass on the basis of the volume of the isolated pore structures measured from electron micrographs. These calculations generally yield only rough estimates from which the approximate number of protomers contained in a pore may be deduced (e.g., Bhakdi and Tranum-Jensen 1985).

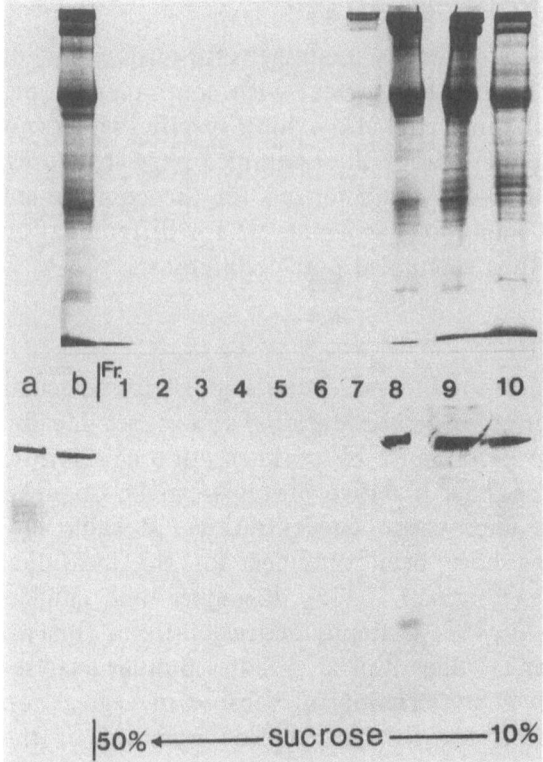

Fig. 3. SDS-PAGE of rabbit erythrocyte membranes lysed with *E. coli* hemolysin (*lane b*) and of fractions recovered after sucrose density gradient centrifugation of DOC-solubilized membranes (*fractions 1–10;* direction of sedimentation if from *right* to *left* as indicated). *Upper panel,* gel stained with Coomassie blue; *lower panel,* immunoblot developed with antibodies to the hemolysin. A sample of *E. coli* hemolysin that was used in the experiment was also applied in the SDS-PAGE immunoblot analysis (*lane a,* lower panel). Note the binding of the hemolysin of M_r 10700 to the erythrocyte membranes and the presence of membrane-derived, detergent-treated toxin exclusively in monomer from in the top three fractions of the gradient (From Bhakdi et al. 1986)

When pore proteins do not form oligomers that can be recovered in detergent, their physical state may be assessed by analyzing sucrose density gradient fractions by SDS-PAGE immunoblotting or by fused rocket immunoelectrophoresis. The use of specific antisera then permits immunological identification of the protein(s) within the impure fractions (Fig. 3).

When assessing the size of membrane-derived pore structures by ultracentrifugation or gel chromatography, consideration must be given to the binding of detergent by these amphiphilic molecules. Thus molecular-weight estimates obtained by ultracentrifugation in detergent

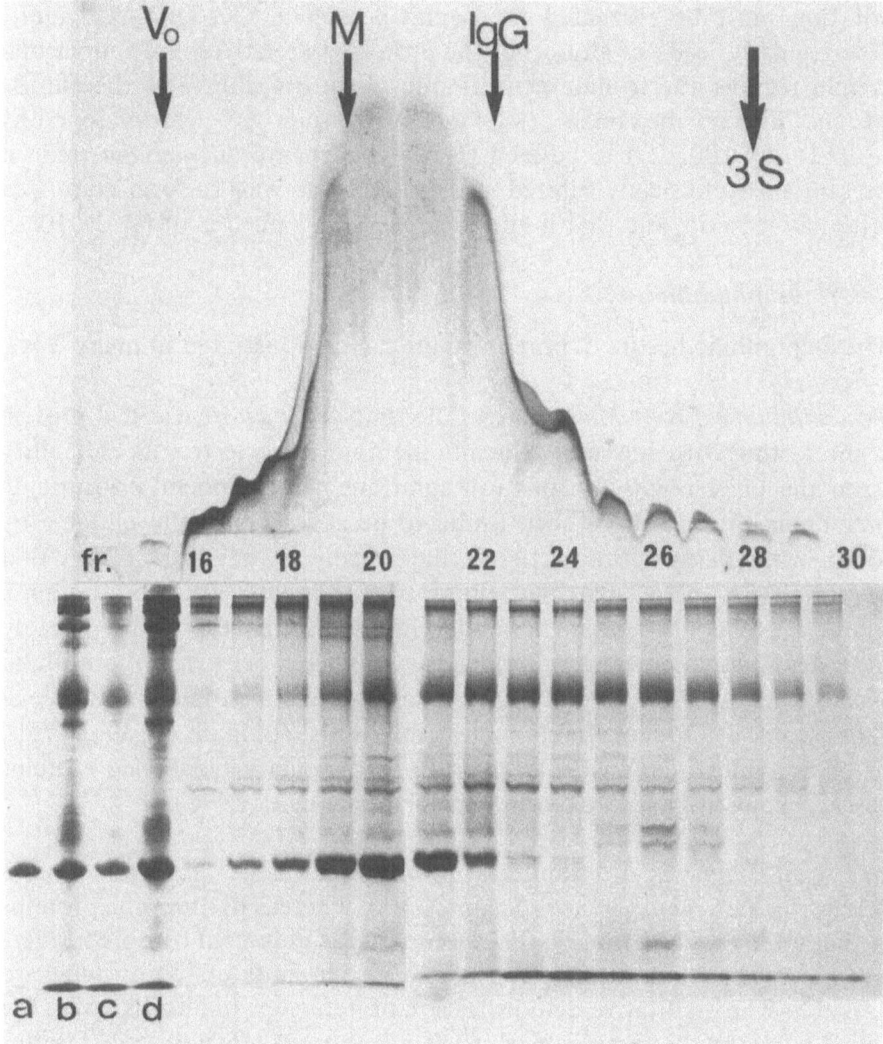

Fig. 4. Elution profile of Triton-extracted, α-toxin-treated rabbit erythrocyte membrane proteins uppon Sephacryl S-300 chromatography, monitored by rocket immunoelectrophoresis and SDS-PAGE. SDS gels *a–d* were loaded with native toxin (*a*), toxin-treated membranes (*b*), Triton-extract of toxin-treated membranes (*c*), and the solubilized Triton membrane residue (*d*). A detergent extract (polypeptide pattern of gel *c*) was chromatographed over Sephacryl S-300 in the presence of 2.4 mM Triton X-100. Aliquots of each column fraction were analyzed by rocket immunoelectrophoresis (*top*) using an antiserum to α-toxin, and by SDS-PAGE (*bottom*). Eluting positions of α$_2$-macroglobulin (*M*), human serum IgG, and native 3S toxin on the same column are indicated; V_0, column void eluting volume (From Füssle et al. 1981)

solution must be corrected for bound detergent. Gel chromatography also regularly leads to elution of the proteins in relatively high molecular weight regions due to detergent binding. Figure 4 illustrates the elution of the α-toxin hexamers ($M_r \approx 2 \times 10^5$) upon gel chromatography in Triton X-100. It can be seen clearly that membrane-derived toxin is present in a physically altered, large molecular weight form compared with native toxin, and that it elutes earlier than a protein of M_r 2×10^5.

3.2.4 Amphiphilicity

The amphiphilic nature of pore proteins can demonstrated in many ways.

Nonelutability From Membranes. The simplest way to test if a protein is interacting with the apolar membrane domains is to test its elutability from the bilayer with various salt solutions. All peripheral erythrocyte membrane proteins have been found to be at least partially elutable by ionic manipulation (Steck 1974). Empirically, it has been found that 1 mM EDTA, pH 8, partially elutes all nonintegral erythrocyte membrane proteins; therefore nonelutability with this extractant reliably indicates that a protein is intimately associated with membrane lipid (Bhakdi et al. 1975a). Elutability is most easily assessed immunologically, e.g., by rocket immunoelectrophoresis (Bhakdi et al. 1975b) or immunoblotting (Bhakdi et al. 1986). Radioactively labelled proteins can also be used in such experiments (Hammer et al. 1975).

Detergent Binding. Amphiphilic proteins bind large amounts of nondenaturing detergents such as Triton X-100, whereas hydrophilic proteins do not (Helenius and Simons 1975). Binding of detergent by pore formers can be assessed quantitatively if sufficient amounts of the proteins are available. The qualitative demonstration of detergent binding is easier, the simplest method being charge-shift electrophoresis (Helenius and Simons 1977) and charge-shift corssed immunoelectrophoresis (Bhakdi et al. 1977). The charge-shift method is based on parallel electrophoretic analyses of proteins in three detergent systems, namely Triton alone, Triton plus DOC, and Triton plus cetyltriethylammonium bromide (CTAB). Amphiphilic proteins binding Triton micelles exhibit an anodal charge shift in the presence of DOC and a cathodal charge shift in the presence of CTAB (Helenius and Simons 1977). Hydrophilic proteins have never been found to exhibit such bidirectional charge shifts.

Lipid Binding. The most direct demonstration of the lipid-binding properties of pores is by membrane reconstitution experiments. The isolated pores are offered lipid in a detergent solution. Removal of the detergent causes lipid aggregation and, if conditions are favorable, liposomes form. Liposome-bound proteins can subsequently be isolated by centrifugation or flotation. It is important to offer sufficient lipid to the protein preparations to obtain proper liposomes and not just gross protein-lipid aggregates (Bhakdi and Tranum-Jensen 1980).

Ultrastructural Characterization. Electron-microscopic analyses have played an important role in the elucidation of several pore structures, namely of complement (Tranum-Jensen et al. 1978; Bhakdi and Tranum-Jensen 1978; Tschopp et al. 1982a, b), α-toxin (Füssle et al. 1981), and streptolysin-O lesions (Bhakdi et al. 1985). Analyses of the pores generated by cytotoxic lymphocytes are currently underway (Dourmashkin et al. 1980; Henkart 1985). Negative staining may be performed on freshly prepared target membranes, as well as on isolated and liposome-incorporated proteins (for review see Tranum-Jensen 1987a). Examination of liposomes can reveal fine details that are otherwise obscured by

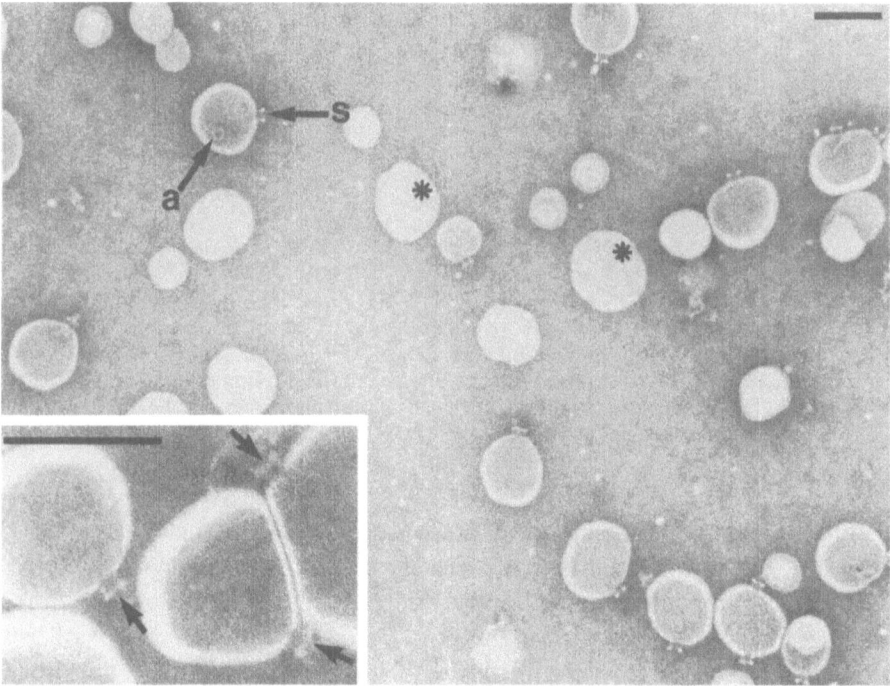

Fig. 5. C5b–9 complement complexes (*arrows*) generated on erythrocytes and subsequently purified and reincorporated into liposomes of phosphatidylcholine. Vesicles that escaped incorporation of a complex (*asterisks*) are characteristically impermeable to the stain. Typically, the complexes project 10 nm exterior to the plane of the membrane (From Bhakdi and Tranum-Jensen 1984a)

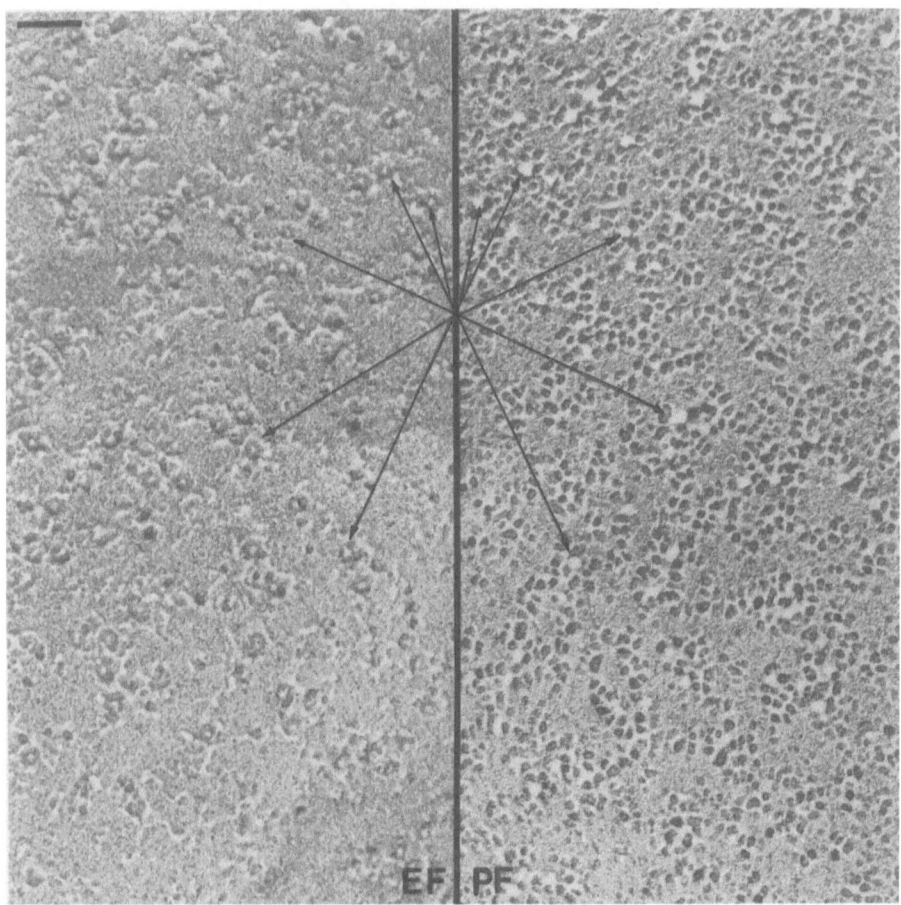

Fig. 6. Complementary freeze-etch replicas of an antibody-sensitized sheep erythro-cyte, lysed with human complement. EF faces (*left*) exhibit numerous ring-shaped structures, interpreted to represent the intramembranous portion of C5b–9 cylindrical complexes. The rings are complementary to circular defects in the lipid plateau of the inner membrane leaflet (PF-face) (For details see Tranum-Jensen and Bhakdi 1983)

membrane constituents. Careful preparation and staining of liposomes generally permits leakiness to the negative stain to be assessed directly. Liposomes carrying pores will appear dark due to the presence of stain, whereas liposomes not carrying a pore will appear light because of the absence of stain (Fig. 5). Additionally, interruptions in the continuity of the lipid bilayer may be seen if lesions are sufficiently large.

Freeze-etch electron microscopy permits visualization of the intra-membraneous domains of large pores (Tranum-Jensen, 1987b) and the demonstration of pore structures on the external membrane surface is generally not problematic. Upon freeze-fracturing, oligomerized pore structures usually partition to follow the external monolayer and can be visualized on the exoplasmic face (EF), leaving holes on the complementary fracture face (proteoplasmic face, PF). The demonstration of the extracted lesions on the face and the holes on the PF face requires special care in preparation of the replicas. The best, albeit most demanding, method of examining the three-dimensional structure of oligomeric pores is to prepare complementary fractures. Such studies yield conclusive evidence for the pore structure of the complement lesion (Fig. 6).

4 Functional Characterization of Pores

4.1 Cells

Erythrocytes are the simplest cells to handle. They may be examined intact or as resealed ghosts. A reliable indication of the existence of pores with effective diameters $<4-5$ nm can be obtained by osmotic protection experiments. Erythrocytes are suspended in isotonic salt solutions supplemented with appropriate concentrations of inert, uncharged molecules of different sizes (e.g., sucrose, raffinose, dextrans). If the protectant molecule cannot pass the pores, osmotic swelling and hemolysis is completely inhibited since the colloid osmotic pressure of intracellular hemoglobin is balanced by the extracellular osmotic protectant molecule. The occurrence of hemolysis indicates that the pore is larger than the protectant molecules. Osmotic protection experiments were first used in the study of complement, where they yielded early evidence for the pore character of the lesions (Green et al. 1959). They have subsequently been used for characterization of the α-toxin (Bhakdi et al. 1984a) and *E. coli* hemolysin pores (Bhakdi et al. 1986).

Another simple approach is to measure the influx of a marker into erythrocytes and the efflux of K^+. Such experiments are best performed under osmotic protection when this is possible. A radioactive marker is added to a dense, osmotically protected cell suspension. Aliquots are removed at appropriate intervals after the addition of the pore former, and radioactivity is measured in the supernatant. Marker influx is simply reflected by a decrease in extracellular radioactivity. The selection of suitable marker molecules for which no active membrane transport system exists and which do not interact with membrane surface components permits straightforward interpretation (Fig. 7; Jorgensen et al. 1983; Bhakdi et al. 1986).

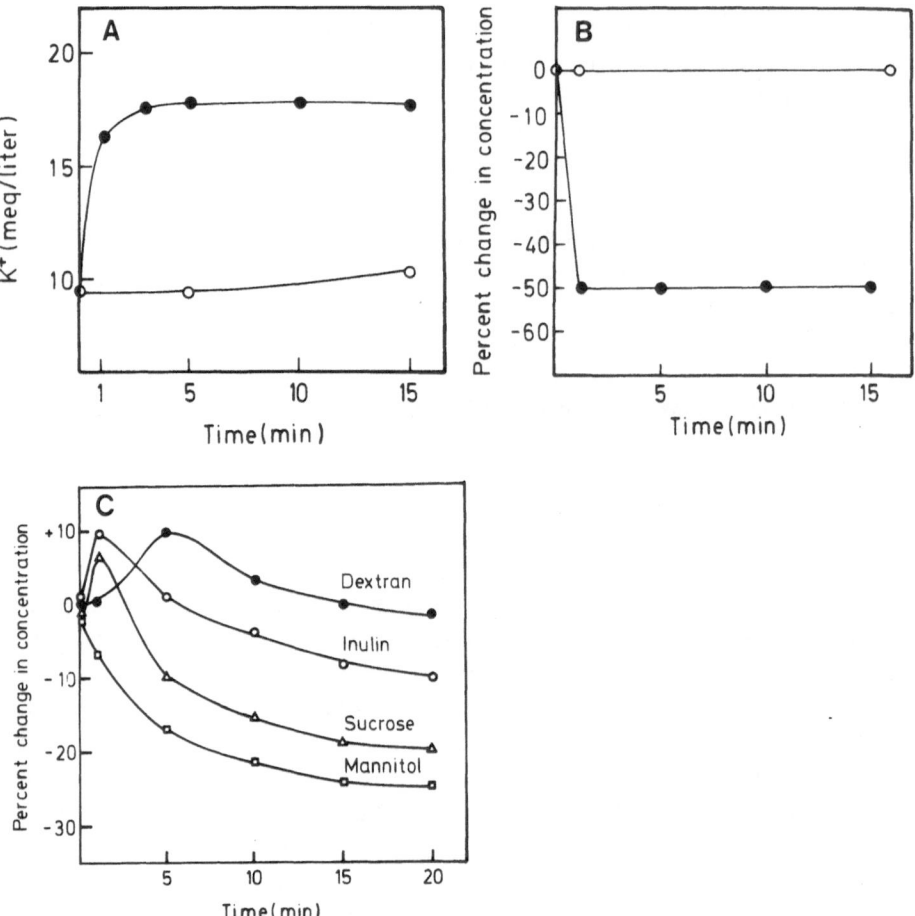

Fig. 7. A Efflux of K^+ from rabbit erythrocytes induced by *E. coli* hemolysin. Erythrocytes were suspended in saline containing 30 mM dextran 4 (2×10^9 cells/ml). Toxin treatment induced a rapid efflux of K^+ that was essentially complete within 1 min. Cell lysis was totally inhibited by the dextran throughout the duration of the experiment. *Symbols; filled circles,* toxin-treated cells; *open circles,* control cells. **B** Toxin-induced influx of $^{45}Ca^{2+}$ into dextran-protected erythrocytes. A suspension of erythrocytes in dextran-containing buffer was incubated with $^{45}Ca^{2+}$. After toxin treatment, samples were removed and radioactivity in the supernatants was measured. Influx of $^{45}Ca^{2+}$ was reflected by a decrease in radioactivity, expressed as percentage change in $^{45}Ca^{2+}$ concentration. Intracellular accumulation of $^{45}Ca^{2+}$ was enhanced due to Ca^{2+}-binding proteins in the cytoplasm. Symbols as in A. **C** Influx of radioactive markers into toxin-treated erythrocytes. The experimental design was as in B (From Bhakdi et al. 1986, with permission)

The assessment of the kinetics of radioactive marker influx or release from resealed erythrocyte ghosts is another, albeit methodologically more demanding, way of demonstrating the presence of transmembrane pores. This approach has been extensively used in the study of complement (Giavedoni et al. 1979; Ramm et al. 1982a, b, 1985). Sophisticated analyses have led to the detection of heterogenoues pores, apparently reflecting the presence of complement complexes containing varying numbers of C9 molecules (Ramm et al. 1985). Marker-release studies have also been used to demonstrate pores generated by α-toxin (Füssle et al. 1981), streptolysin-O (Buckingham and Duncan 1982), *Pseudomonas aeruginosa* cytolysin (Lutz et al. 1982), and pores caused by cytotoxic lymphocytes (Simone and Henkart 1980).

A method of corroborating these functional studies is to demonstrate specific inhibition of the flux by treatment with antibodies. This approach has been used with apparent success to demonstrate inhibition of marker flux through the complement lesion by C9-specific monoclonal antibodies (Morgan et al. 1984) and C5-specific polyclonal antibodies (Simone and Henkart 1982). Evidence obtained with C9-specific monoclonal antibodies has indicated that an epitope of C9 becomes exposed at the cytoplasmic surface of the erythrocyte membrane (Morgan et al. 1984; however, see comment on p. 188).

When nucleated cells are used as targets, the possibility of pore removal and lesion repair occurring in parallel with membrane attack must be considered. Thus, higher doses of the pore formers may be needed to detect functional lesions and the overall effective pore size may appear smaller. Functional studies in nucleated cells have included measurements of marker release and of marker influx with α-toxin (Thelestam et al. 1973; Thelestam and Möllby 1975; Suttorp et al. 1985a; Ahnert-Hilger et al. 1985), complement (Koski et al. 1983; Seeger et al. 1986) and *P. aeruginosa* cytolysin (Lutz et al. 1982; Suttorp et al. 1985b). The data have all conformed to the pore concept of action of these proteins in nucleated cells.

4.2 Artificial Lipid Bilayers

Liposomes and black-lipid membranes are useful model systems to study pore formation. Liposomes were used in early studies of complement (Kinsky 1972; Lachmann et al. 1970) and α-toxin (Arbuthnott et al. 1973). In both cases release of liposome-entrapped markers occurred, indicating overt damage to the bilayers. Planar lipid membranes have been used in studies of complement proteins (Michaels et al. 1976), α-toxin (Menestrina 1986), lymphocyte pores (Henkart and Blumenthal

1975, Young et al. 1986a, b), *E. coli* hemolysin (G. Menestrina et al., unpublished work) and cytolysins from sea anemone (Michaels 1979) and amoeba (Lynch et al. 1982, Young et al. 1982). Increases in electrical conductance across the bilayers permit estimates to be made of the size, life-span and homogeneity of the pores. Details of the methods are available in several reviews (e.g., Montal 1974).

5 Pores with Recognizable Ultrastructure

5.1 Staphylococcal α-Toxin

5.5.1 Biological Significance

α-Toxin is produced by most strains of *Staphylococcus aureus* and is considered a major determinant of staphylococcal pathogenicity (McCartney and Arbuthnott 1978). The toxin attacks all mammalian cells studied, albeit with greatly varying efficacy. It is hemolytic, cytotoxic, and lethal when injected in large amounts. Antitoxin antibodies are found in the plasma of all healthy individuals and it is thus apparent that the toxin is produced by staphylococci in human hosts. Staphylococcal strains that do not produce α-toxin have been reported to be less virulent in animal experiments. The general consensus is that the toxin represents a significant factor in staphylococcal pathogenicity.

5.1.2 Properties of Native Toxin

The toxin is secreted as a water-soluble, single polypeptide chain devoid of amino sugars with M_r 34 000 (Bhakdi et al. 1981; Gray and Kehoe 1984). The protein has an $S_{20,w}$ (sedimentation coefficient in water at 20° C) of 3.3S, and an isoelectric point (pI) of 8.6. The primary amino acid sequence has been published (Gray and Kehoe 1984), but no unusual conformations can be deduced. The protein exhibits 6% α-helical conformation and 68% β-sheet structure (Tobkes et al. 1985). It spontaneously aggregates to form hexamers under specific heat conditions (Arbuthnott et al. 1967) and is inactivated by the binding of specific antibody as well as by binding to plasma low-density lipoprotein (LDL) (Bhakdi et al. 1983a). The native protein is degraded and destroyed by proteases.

5.1.3 Primary Mode of Toxin Binding to Membranes and the Question of Specific Binder Molecules

Two major findings have constituted a puzzle in this area. On the one hand, the toxin binds to liposomes devoid of proteins (Freer et al. 1968), to planar-lipid membrane composed of pure phosphatidylcholine (Menestrina 1986), to plasma LDL (Bhakdi et al. 1983a), and to all mammalian cell membranes studied to date (McCartney and Arbuthnott 1978). Binding to these substrates is of widely varying efficiency, e.g., it is low in the case of erythrocytes, but high in the case of LDL (Fig. 8). These findings indicate that α-toxin binds in a rather nondiscriminating fashion to lipids in general, without requiring a specific binder molecule. On the other hand, different cell targets display a very wide spectrum of sensitivity towards the toxin, e.g., rabbit erythrocytes are approximately 100-fold more sensitive than human erythrocytes (McCartney and Arbuthnott 1978; Rogolsky 1979; Harshman 1979; Bernheimer 1974). This finding in turn suggests that rabbit cells may possess binder molecules of high affinity for the toxin, or that certain cells possess a means of defence against insertion of the toxin in the membrane. Evidence for increased binding affinity was obtained in analyses using radio-iodinated toxin, which suggested the presence of a few specific binder molecules on rabbit erythrocytes that were absent on human cells (Cassidy et al. 1976). A consensus view would be that specific "receptors" are not required for toxin binding, but that certain cells possess high-affinity binders that render them highly sensitive to the toxin. The possibility that such mambrane components could cause toxin accumulation in certain tissues in vivo is worth considering.

5.1.4 Pore Formation by α-Toxin

The first detailed investigations into the mode of membrane damage by α-toxin were performed by Arbuthnott et al. (1967, 1973). These investigators discovered that the toxin would attack and perturb liposomal membranes (Freeret al. 1968). They reported that ring structures, possibly representing toxin hexamers, could by seen by electron-microscopic examination of attacked membranes (Freer et al. 1973), and noted that formation of ring structures could also be induced by subjecting purified toxin preparations to specific heat conditions (Arbuthnott et al. 1967). Hence, the ring structures were correctly deduced to represent oligomeric aggregates of the toxin. Further studies by Buckelew and Colacicco (1971) demonstrated that α-toxin would perturb lipid monolayers and Arbuthnott et al. (1973) concluded that α-toxin must damage membranes through an intimate association and interaction with the membrane lipids. Thelestam et al. (1973) and Thelestam and Möllby

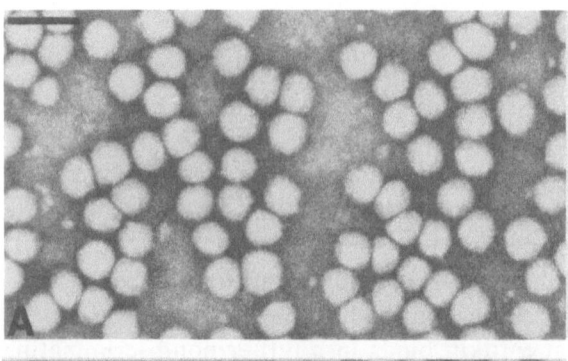
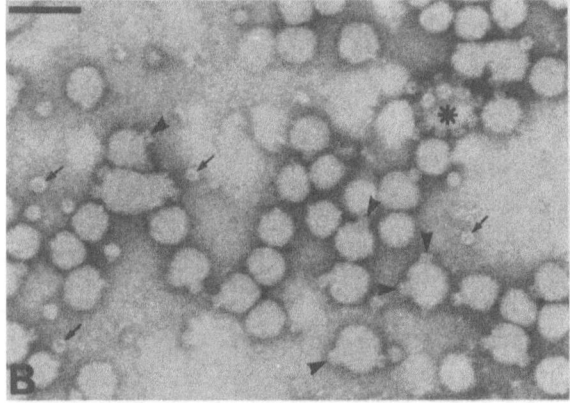
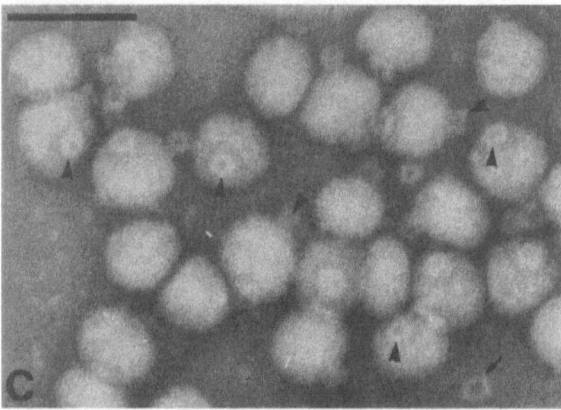
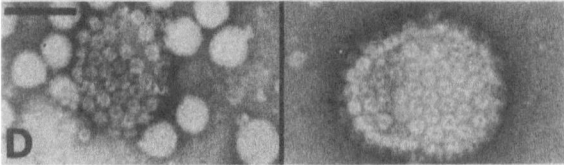

Fig. 8 A—D. Binding of *S. aureus* α-toxin to human plasma low-density lipoprotein (LDL). **A** Control preparation of human LDL showing a uniform population of smooth and round, approximately 25 nm particles; **B, C** toxin-treated LDL. Typical ring and stub profiles of 11S α-toxin hexamers are seen attached to the LDL particles (*arrow heads*). "Free" 11S toxin complexes are indicated by *small arrows;* toxin detachment from LDL possibly occurred during the negative staining procedure. A minor fraction of the LDL particles carry markedly more toxin structures than others (*asterisk*). **D** A small fraction of large 100—150 nm particles, possibly of a composition different from the 25 nm particles, are heavily loaded with 11S toxin structures. Sodium silicotungstate negative staining. Scale bars indicate 50 nm (From Bhakdi et al. 1983a)

(1975) contributed to the understanding of α-toxin action by showing that nucleated cells treated with the purified toxin release low molecular weight, intracellular markers and retain cytoplasmic proteins. These studies were performed at a time when the pore concept, just formulated in the study of complement (Mayer 1972), had not yet been extended to other membrane-damaging proteins. Following the isolation of the complement pore (Bhakdi and Tranum-Jensen 1978), it became obvious that the α-toxin rings most probably represented a counterpart of the complement lesion. This was borne out by the isolation of the α-toxin pores and the demonstration of their amphiphilic nature (Füssle et al. 1981; Fig. 9). Functional studies using erythrocyte targets indicated that the toxin generated pores of around 2 nm diameter, and penetration of α-toxin into the hydrophobic membrane domain was also demonstrated by labeling with a hydrophobic membrane probe (Thelestam et al. 1983). The hexamer nature of the toxin pores was established by determination of their molecular weights using ultracentrifugation analyses (Bhakdi et al. 1981). Biochemical cross-linking studies have fully supported these findings (Tobkes et al. 1985). Hence, the concept that α-toxin damages membranes by hexamerization to form ring-structured pores is supported by many lines of evidence and is today widely accepted.

5.1.5 α-toxin and Ca^{2+}

Extracellular Ca^{2+} appears to influence both the primary event of toxin pore formation and secondary intracellular processes. Millimolar concentrations (5−30 mM) of extracellular Ca^{2+} inhibit pore formation, apparently without inhibiting the primary binding of toxin to the membrane (Bashford et al. 1984; Harshman and Sugg 1985). It has been proposed that this is due to inhibition of lateral movement and aggregation of the toxin molecules in the membrane (Harshman and Sugg 1985), but studies in planar-lipid membranes suggest that Ca^{2+} more probably acts on the formed hexamers (Menestrina 1986). The biological significance of Ca^{2+} inhibition of the α-toxin pore requires clarification.

Extracellular Ca^{2+} also appears to be responsible for triggering important secondary events in nucleated cells treated with subcytolytic doses of toxin. Current evidence indicates that influx of Ca^{2+}, probably through the toxin pores, will, for example, activate the arachidonic acid cascade in endothelial cells (Suttorp et al. 1985a). Generation of arachidonic

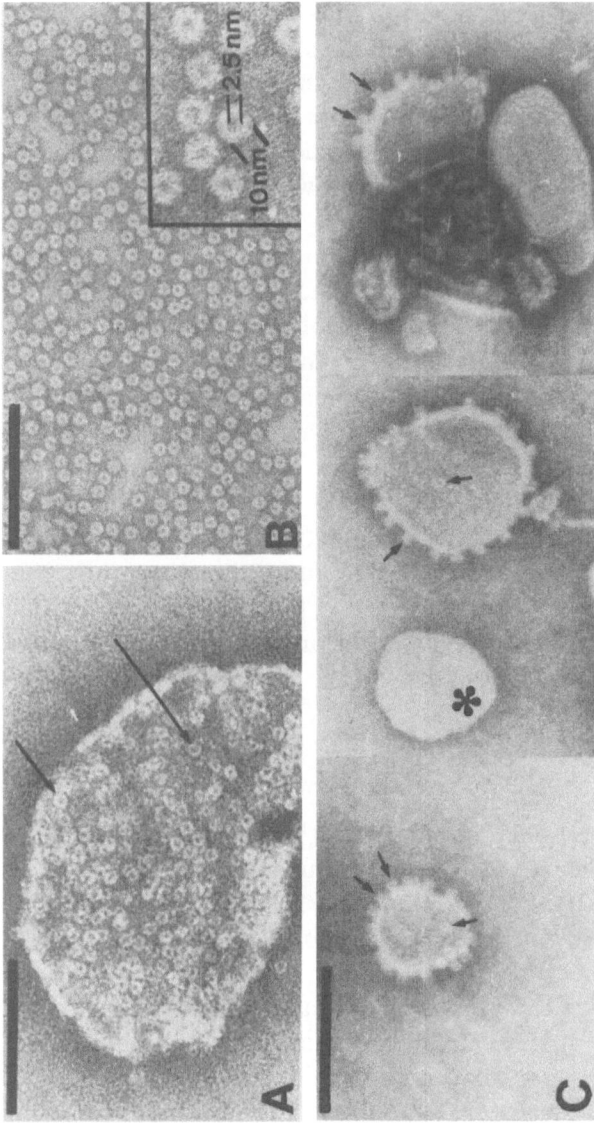

Fig. 9. A Negatively stained fragment of rabbit erythrocyte lysed with *S. aureus* α-toxin. Numerous 10 nm ring-shaped structures are seen over the membrane (*arrows*). **B** Isolated toxin hexamers in detergent solution. **C** Lecithin liposomes carrying reincorporated toxin hexamers that are seen as stubs along the edge of the liposomal membrane and as rings over the membrane (*arrows*). Characteristically, liposomes that escape incorporation of the toxin are impermeable to the stain. Sodium silicotungstate negative staining. Scale bars 100 nm (From Bhakdi and Tranum-Jensen 1984b)

acid metabolites will obviously induce a wide spectrum of reactions, depending on the cell target (Seeger et al. 1984). Such processes may be partially responsible for the inflammatory reactions that are so characteristic of staphylococcal infections (Fig. 10).

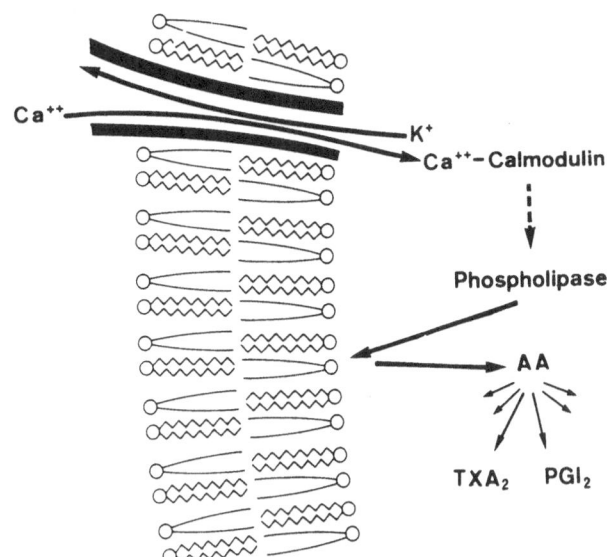

Fig. 10. Proposed mechanism of mode of action of α-toxin. A toxin-created trans-membrane pore serves as a nonphysiological calcium channel. Incoming Ca^{2+}, possibly after binding to calmodulin, activates phospholipases with subsequent cleavage of arachidonic acid and formation of tissue-specific arachidonic acid metabolites. *TXA₂*, thromboxane A_2; *PGI₂*, prostacyclin (From Suttorp et al. 1985a)

5.1.6 α-Toxin as a Membrane-Permeabilizing Agent

Since α-toxin generates relatively small, stable pores in cell membranes, it emerges as a candidate for selective membrane permeabilization. Initial studies in this area have brought forth very promising results with rat hepatocytes (McEwen and Arion 1985) and chromaffin cells (Ahnert-Hilger et al. 1985). In the latter experiments, micromolar concentrations of extracellular Ca^{2+} were found to induce exocytosis and dopamine release in cells treated with subcytolytic doses of toxin (Ahnert-Hilger et al. 1985; Bader et al. 1986). Further studies using α-toxin as a permeabilizing agent may yield new information about the mechanisms involved in exocytotic processes.

5.1.7 Myelin Disruption by α-Toxin

Because of its affinity for diverse types of lipids and apolar substances, α-toxin may exert biological effects by mechanisms other than pore formation. Disruption of myelin by the toxin is one consequence of

possible biological significance (Harshman et al. 1985). It has been reported that α-toxin applied intravenously accumulates in the central
nervous system (Jeljaszewicz et al. 1978), but this claim requires con-
firmation.

5.2 Streptolysin-O

5.2.1 Biological Significance

Streptolysin-O is the prototype of sulfhydryl-activated bacterial cytolysins (Bernheimer 1974; Smyth and Duncan 1978; Alouf 1980). These
comprise a group of at least 15 toxins that are produced by a variety of
gram-positive bacteria, and include listeriolysin (Parrisius et al. 1986),
tetanolysin (Alving et al. 1979), cereolysin (Cowell et al. 1978), and
perfringolysin (Smyth et al. 1975; Mitsui et al. 1979a, b), all of
which have several properties in common. The toxins are produced
as single polypeptide chains with reported M_r ranging from 40 000 to
80 000. They are reversibly inactivated by atmospheric oxygen and
attack only membranes containing cholesterol. Thus, they attack all
mammalian cells but not bacteria. In many cases, membrane damage
by these toxins has been shown to be accompanied by the appearance of
large semicircular and circular structures on the membranes that can be
visualized by electron microscopy (for example, see Bernheimer 1974;
Duncan and Schlegel 1975; Smyth and Duncan 1978; Alouf 1980;
Bernheimer and Rudy 1986).

   Streptolysin-O is produced by β-hemolytic group A streptococci,
which represent the major human pathogens of the genus *Streptococcus*.
Antibodies against the toxin are found in adults and titers may rise
dramatically during and following streptococcal infections, implying that
the toxin is produced by the bacteria in human hosts. Persistently high
anti-streptolysin-O titers are encountered in nonsuppurative poststreptococcal sequelae; in particular, they are invariably observed in all cases of
rheumatic fever. The cause and significance of these high antibody
titers remain unclear.

   Many studies have been performed on the numerous biological
effects of streptolysin-O in vivo (for review see Alouf 1980). The toxin
is lethal in a variety of animal species (Todd 1938; Herbert and Todd
1941; Halbert et al. 1963a, b), primarily because of a marked acute
cardiotoxicity (Halbert et al. 1963a, b; Halpern and Rahman 1968),
but the pathophysiology of this phenomenon is not fully understood
(Alouf 1980). Intravenous injection of streptolysin-O also elicits severe
dysfunction of the central nervous system (Gupta and Srimal 1979).

5.2.2 Properties of Native Streptolysin-O

Several methods for isolating streptolysin-O have been reported (e.g. van Epps and Anderson 1969, 1973; Alouf and Raynaud 1973; Shany et al. 1973; Linder 1979). Purifications are impeded by the presence of proteases cosecreted by the streptococci, and satisfactory yields are therefore difficult to obtain. There were somewhat puzzling discrepancies between earlier reports of the molecular weight and isoelectric point of the toxin. A consensus was reached when it was found that two hemolytic forms of streptolysin-O could be recovered; truly native toxin possessed a M_r of 69 000 (P^I 6.0–6.4); this molecule may be nicked and partially degraded to a hemolytically active molecule of M_r 57 000 (P^I 7.0–7.5) (Bhakdi et al. 1984b). The biological significance and role of the removed portion of M_r 10 000–12 000 is unknown.

Streptolysin-O exhibits potent hemolytic and cytolytic activities; hemolytic titrations indicate that the presence of approximately 100 molecules of the toxin induces the formation of one functional lesion per erythrocyte (Alouf 1980; Bhakdi et al. 1984b). In contrast to α-toxin, no marked between erythrocytes of different species. Molecular cloning of the streptolysin-O gene in *E. coli* has been reported (Kehoe and Timmis 1984).

5.2.3 Interaction of Streptolysin-O with Cholesterol

The most extensive studies on the interaction of sulfhydryl-activated cytolysins with cholesterol have been performed with streptolyin-O (for review see Alouf 1980). The key findings have been that pure cholesterol inhibits (irreversibly inactivates) streptolysin-O (Hewitt and Todd 1939; Bernheimer 1974; Duncan and Schlegel 1975; Smyth and Duncan 1978; Alouf 1980) and causes the formation of the characteristic oligomeric toxin structures (Duncan and Schlegel 1975). Studies on the stereo-specificity of the sterol effect (reviewed by Alouf 1980) indicate requirements for the 3β-OH group (Howard et al. 1953; Alouf and Geoffroy 1979), a lateral aliphatic side chain of suitable size at carbon-17 (Prigent and Alouf 1976; Watson and Kerr 1974), and an intact B-ring (Alouf 1980). Essentially the same structural requirements are given for the polyene antibiotics (Norman et al. 1976).

That sulfhydryl-activated toxins form complexes with cholesterol has been inferred form several experiments. These include precipitation in agar gels (Prigent and Alouf 1976; Alouf and Geofoy 1979), and electron-microscopic studies showing formation of arc and ring structures upon incubation of toxin with cholesterol or cholesterol-containing membranes (e.g. Duncan and Schlegel 1975; Smyth et al. 1975; Cowell et al.

1978; Mitsui et al. 1979b; Rottem et al. 1982). However, there has been only one more detailed quantitative study of cholesterol binding by a cytolysin of this group (Johnson et al. 1980). This study sought directly to demonstrate binding of radiolabeled cholesterol to purified cytolysin. Toxin preparations were incubated whith cholesterol solutions at concentrations in which the cholesterol was present in soluble, micellar form. The study directly demonstrated binding of such micellar cholesterol to streptolysin-O and also showed that the toxin-cholesterol complex was hemolytically inactive and in macromolecular form. Technical difficulties did not permit experiments to be performed at lower cholesterol concentrations where the lipid would be present in monomer form. Overall, the findings clearly show that cholesterol plays a key role in inducing the oligomerization of streptolysin-O. Nevertheless, this statement should not yet be equated with the statement that cholesterol is the specific membrane receptor for the toxin, and studies of the binding of submicellar concentrations of cholesterol to monomeric streptolysin-O still need to be performed. Maintenance of the toxin in its monomeric form during the experiment appears important since apolar binding sites will probably become exposed whenever toxin oligomerization occurs; these will secondarily bind further lipid molecules and thus hamper clear interpretation of any binding data. Such studies have yet to be undertaken.

5.2.4 Pore Formation by Streptolysin-O

Dourmashkin and Rosse (1966) first reported the presence of circular and semicircular lesions on erythrocyte membranes after treatment with streptolysin-O. In a detailed investigation, Duncan and Schlegel (1975) established a causal link between formation of these structures and membrane damage, and showed that the toxin disrupted the integrity of liposomes. A finding of major significance emerging from these studies was that streptolysin-O probably entered into an intimate association with membrane lipid to generate functional lesions. Until recently, the general assumption was that sequestration of lipids occurred through binding of the toxin to cholesterol, and that this sequestration was the cause of membrane damage. The ultrastructural lesions were widely thought to represent aggregates of the toxin with membrane cholesterol. Functional studies performed by Duncan and colleagues (Duncan 1974; Buckingham and Duncan 1983) showed that treatment of cells with streptolysin-O caused rapid release of even very large intracellular markers; hence, the toxin lesions appeared to exhibit effective diameters exceeding 15 nm. That the toxin might form such large pores through embeding of oligomeric structures in the bilayer was, however, deemed unlikely by most investigators, mainly because lesions could not be demonstrated on the inner leaflets of cell membranes by freeze-fracture electron microscopy (Smyth and Duncan 1978; Niedermayer 1985).

This issue was resolved recently when streptolysin-O oligomers were isolated from target membranes and identified as amphiphilic protein pore structures (Bhakdi et al. 1985). Lipid was not detectable in these protein preparations; hence, cholesterol, although responsible for initial binding of the toxin to the target membrane, obviously did not significantly contribute to the formation of the pore structure. Further because streptolysin-O oligomers could be reconstituted into bilayers of lecithin, the apolar regions of the oligomeric complex clearly did not display any specific requirement for cholesterol to enable its insertion, once formed. Electron-microscopic examination of recon-stituted liposomes was particularly revealing since it permitted direct visualization of the oligomeric protein embedded within the bilayer, and also demonstrated the presence of very large transmembrane pores (Fig. 11). Final corroboration of the pore concept was obtained by freeze-fracture electron microscopy. Low-angle rotational shadowing, as previously applied to study the complement lesion, revealed defects in the cytoplasmic lipid monolayer (Bhakdi et al. 1985). Semicircular lesions appeared bound by an edge of free lipid. In conjunction with observations obtained by negative-staining electron microscopy of reconstituted liposomes, these results raise the intriguing possibility that transmembrane pores may be generated by insertion of non-circular protein structures, with membrane lipid alone completing the circum-ference of the lesion. It has been proposed that the hydrophilic sides of the protein may repel lipid molecules in the manner shown in Fig. 12. If this hypothesis is confirmed, the model could be extended to account for pore formation by monomeric protein molecules, as well as by native, physiological pores.

The molecular weights of toxin oligomers have only been estimated by determination of their volumes by electron microscopy. These calculations indicated that between 25 and 80 monomers participate in forming partially of fully circular lesions. It is conceivable that lesions containing low numbers of protomers exist that are not clearly re-cognizable by electron microscopy (Bhakdi et al. 1985).

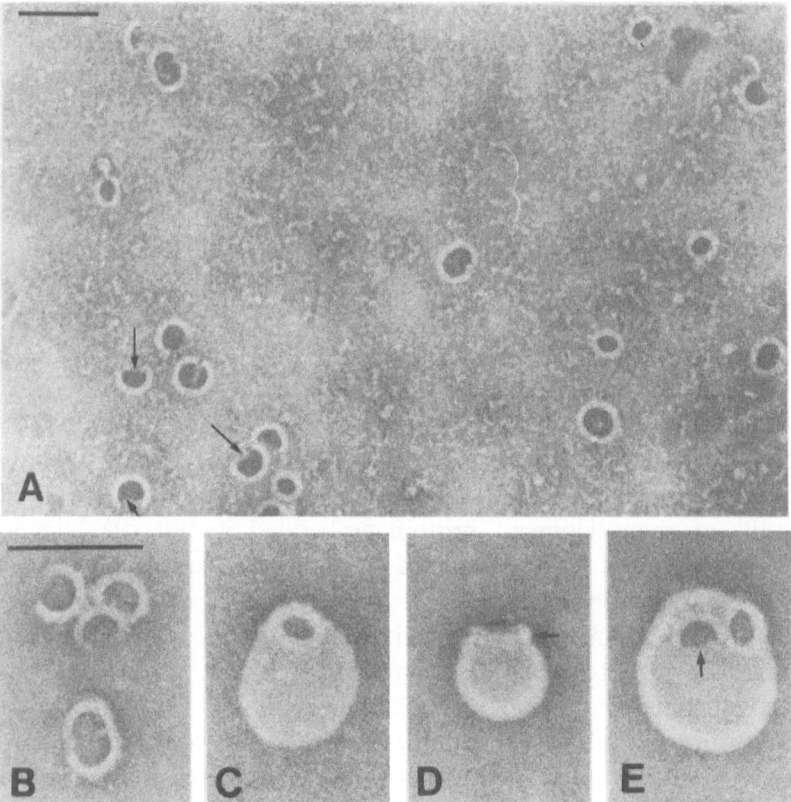

Fig. 11. A Negatively stained erythrocyte membrane lysed by streptolysin-O showing numerous 25–100 nm long approximately 7.5 nm broad, curved rods of 13–16 nm inner radius of curvature. Most rods are approximately semicircular, often joined in pairs at their ends. Dense accumulations of stain are seen at the concave side of the rods. When these do not form closed profiles, the stain deposit is partly bordered by a "free" edge of the erythrocyte membrane (*arrows*). **B** Negative staining of isolated streptolysin-O oligomers, showing curved rod structures identical to those found on the membranes. **C–E** Purified streptolysin-O oligomers reincorporated into cholesterol-free lecithin liposomes. The toxin oligomers form holes in the liposomal membranes. Part of the circumference of such holes appears bordered by a free edge of liposomal membrane in many instances (*arrows*). Scale bars 100 nm. Sodium silicotungstate negative staining (For details see Bhakdi et al. 1985)

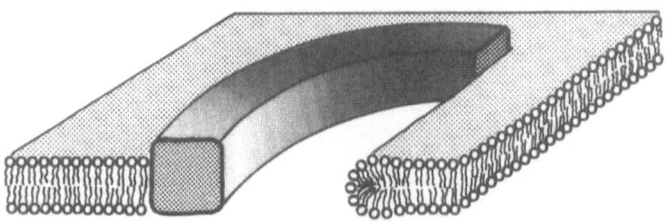

Fig. 12. Diagrammatic presentation of the gross, principle features proposed for the streptolysin-O-induced membrane lesion generated by incompletely circular toxin oligomers (From Bhakdi et al. 1985)

5.2.5 Mode of Pore Formation

Streptolysin-O binds to erythrocytes at $0°$ C but hemolysis does not ensue at low temperature unless the number of bound toxin molecules exceeds $5000-10\,000$ per cell (Alouf 1980; Hugo et al. 1986). When cells are treated with subhemolytic doses at $0°$ C, washed, and resuspended in toxin-free buffer at $37°$ C, lysis occurs. These experiments indicate that binding of toxin to the membrane can be partially dissociated from the process of transmembrane pore formation (Alouf 1980). In an extension of these studies, a neutralizing monoclonal antibody was used to probe the mode of transmembrane pore formation by streptolysin-O. It was found that the antibody did not prevent toxin binding, but totally suppressed toxin oligomerization. Thus, the membrane-damaging function of streptolysin-O was related to the process of oligomer formation in the membrane (Hugo et al. 1986). Further, it became apparent that streptolysin-O molecules first bind to the membrane, oligomerization taking place in a second step through lateral aggregation in the bilayer. Additional analyses indicated that the latter event causes no release of lipid from the membranes; hence, pore formation probably involves lateral "forcing aside" of membrane components (S. Bhakdi and J. Tranum-Jensen, unpublished work). If streptolysin-O indeed binds specifically to cholesterol molecules, labeled complexes of the toxin with the neutralizing monoclonal antibody could become a tool to probe the distribution, orientation, and mobility of cholesterol molecules in biological membranes.

5.2.6 Secondary Effects Elicited by Membrane Attack by Streptolysin-O

Experiments in vitro indicate that streptolysin-O and other sulfhydryl-activated cytolysins may exert biological effects at subcytolytic doses. In particular, streptolysin-O appears to elicit inhibitory effects on leucocytes (van Epps and Anderson 1974; Wilkinson 1975), lymphocytes (Anderson and Cone 1974; Anderson and Amirault 1976), and macrophages (Ofek et al. 1972). Subcytolytic doses of streptolysin-O have been shown to cause transient impairment of membrane transport systems (Duncan and Buckingham 1977) and stimulation of the arachidonic acid cascade in human polymorphonuclear granulocytes has also been reported (Bremm et al. 1985).

Attack on cell membranes by streptolysin-O may also perpetuate local inflammation and tissue damage through an entirely different mechanism. We have found that immune complexes formed between streptolysin-O and human IgG exhibit a remarkable capacity for activat-

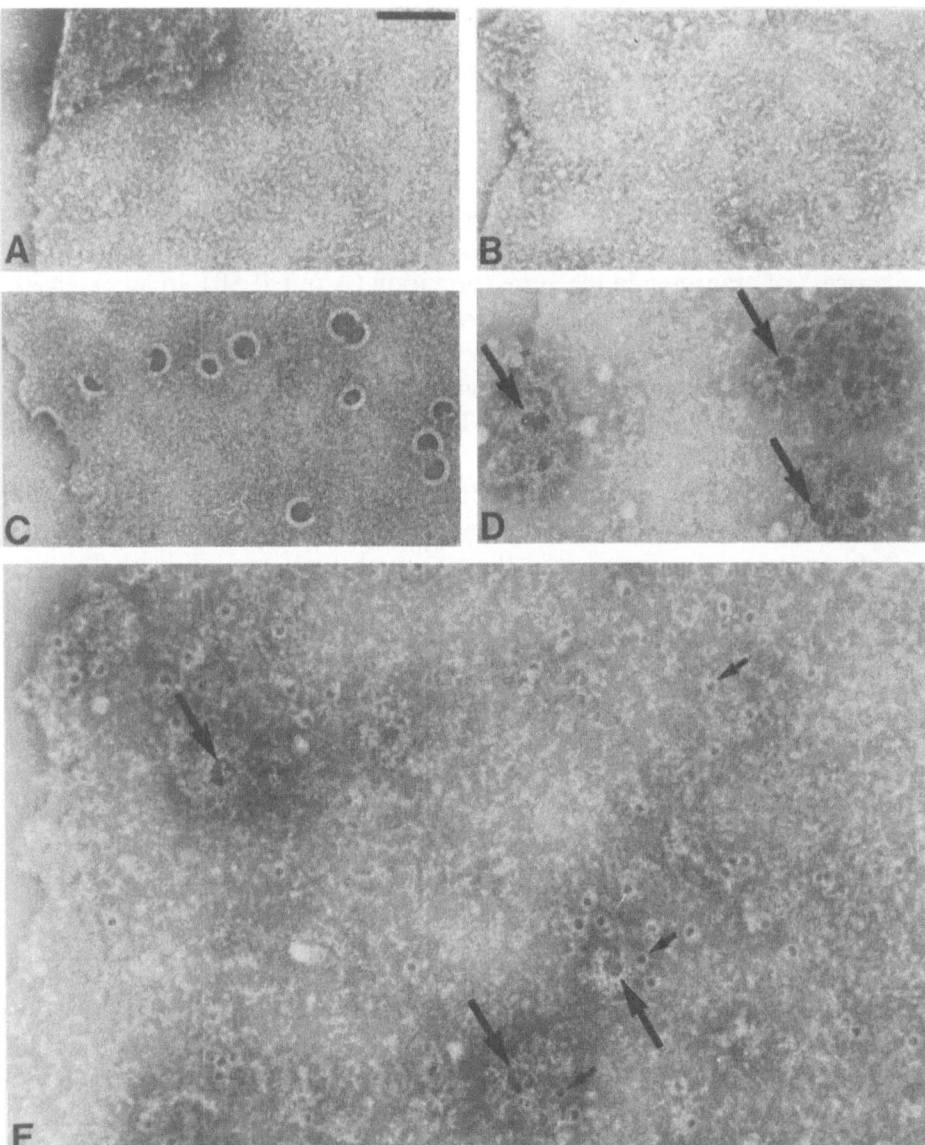

Fig. 13. **A** Human erythrocyte membrane treated with 5 mM dithiothreitol (DTT).
Two layers of membrane are superimposed at *upper left*. **B** DTT-treated membrane
after incubation in autologous human serum. Only a slight coarsening of the surface
results from this treatment. **C** Membrane treated with streptolysin-O; the characteristic
streptolysin-O lesions are seen. **D** Same membrane preparation as in C, posttreated
with autologous serum. The toxin structures have become covered by irregular
tufts of protein like material that extend onto the surrounding membrane. **E** As D,
posttreatment with a high serum dose. Classical complement lesions (*small arrows*)
are scattered over the membrane with a tendency to cluster around the streptolysin-O
lesions (*large arrows*). Scale bar 100 nm (From Bhakdi and Tranum-Jensen 1985)

ing the autologous complement system. When occurring in the fluid phase, this reaction causes complement consumption and may serve to deviate complement attack away from the bacteria. When strepto-lysin-O enters a cell membrane, it generates a hyperactive focus for complement activation; this proceeds to completion with the generation of complement lesions on the autologous membranes (Bhakdi et al. 1985; Fig. 13). Complement activation invariably generates biologically active peptides which may be partially responsible for the inflammatory reactions encountered in streptococcal lesions. Moreover, complement activation on autologous tissue may play a role in the pathogenesis of poststreptococcal disease. Other bacterial products with similar complement-activating properties will probably be identified in the future and it will be of interest to delineate their participation in the genesis of immune-complex diseases in general.

5.3 Complement

5.3.1 Biological Significance

The complement system comprises a group of at least 20 plasma proteins and represents the major "non-specific" humoral defence system in mammals. Many reviews are available of the biochemistry and function of complement components (for example see Müller-Eberhard 1975; Hügli and Müller-Eberhard 1978; Porter and Reid 1978; Kazatchkine and Nydegger 1982), and we will restrict the present discussion to the question of transmembrane pore formation occurring as the terminal step in the complement cascade.

Complement activation can occur either via the classical pathway (involving C1, C4, and C2), or via the alternative pathway which circumvents these components but instead requires the alternative pathway components B, D, and P. Both pathways merge at the level of C3, which is cleaved to yield C3a and C3b. The latter participates in forming a C5-cleaving enzyme (C5 convertase), either through forming a complex with C4, 2, or through interaction with factors B, D, and P. C5 cleavage initiates the terminal reaction pathway by generation of C5a and C5b. The latter molecule complexes with C6–C9 on a cell membrane to form the cytolytic, pore-forming C5–9 (m) complex (Mayer et al. 1981; Bhakdi and Tranum-Jensen 1983; Müller-Eberhard 1984). C5b–9 complex formation can also occur in the fluid phase (i.e., in plasma or serum), in which case an additional plasma protein, the "S-protein", becomes incorporated into the complex to generate SC5b–9 (Kolb and Müller-Eberhard 1975; Bhakdi and Roth 1981). The latter is a cytolytically inactive, hydrophilic macromolecule of unknown biological function. In contrast, membrane-bound C5b—9 complexes generate stable transmembrane pores of fairly large size and are thus endowed with cytolytic and bactericidal properties (Fig. 14). In addition, the C5b—9 complex can inactivate some viruses and kill certain higher parasites. Regardless of the mode of complement activation and the nature of the target under attack, generation of the biologically active peptides (mainly C3a and C5a) causes local inflammatory tissue reactions (Hügli and Müller-Eberhard 1978). Of major biological importance is the strong opsonizing effect of particle-bound C3b (Gigli and Nelson 1968; Ehlenberger and Nussenzweig 1977; Fearon et al. 1981).

The complement system is in a state of dynamic equilibrium, with a continuous series of activating events counterbalanced by effective regulatory, inhibitory mechanisms that halt the cascade at various stages. Regulation is known to occur at virtually every enzymatic step up to C5 cleavage (for example see Müller-Eberhard 1975; Fearon 1979, 1980; Kazatchkine and Nydegger 1982). Therefore, under normal circumstances C5 activation and generation of cytolytic C5b—9 complexes occurs selectively on cell targets marked for attack, e.g., through the binding of antibody or by the binding of alternative-pathway com-

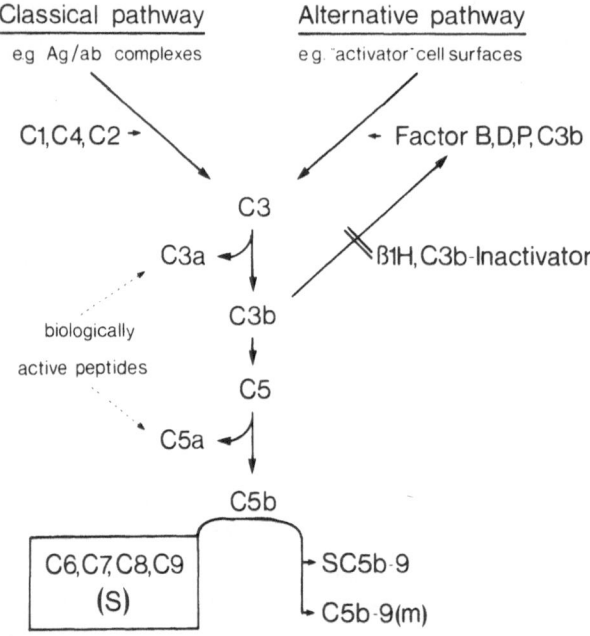

Fig. 14. Schematic representation of the major reaction pathways of the complement system

ponents. It is clearly undesirable that C5b-9 should be generated on normal, autologous cells, and this generally does not happen. However, derangement of the immune recognition system or failure of the regulatory machinery may cause autologous attack by complement. Hence, the biological significance of C5b-9 complexes lies not only in their role as effectors of cytolysis for immune defence, but also in that they may assume deceitful roles in the development of several diseases, including those involving autoimmune mechanisms.

5.3.2 Properties of Native C5-C9 components and the Process of C5b-9 Assembly

The biochemical characteristics of native terminal complement components are summarized in Table 1. All are soluble glycoproteins of M_r 70 000–200 000 (see Hammer et al. 1981). The amino acid sequence of C9 is known (DiScipio et al. 1984; Stanley et al. 1985). This component is cleaved by thrombin to yield relatively hydrophilic (C9a) and hydrophobic fragments (C9b) (Biesecker et al. 1982a), the latter being selectively labeled by apolar photolabels in target lipid bilayers (Ishida et al. 1982; Amiguet et al. 1985). Thrombin-nicked C9 fully retains its pore-forming function without generating ultrastructural circular lesions (Dankert and Esser 1985).

The terminal pathway is initiated when C5 is cleaved to C5b, optimally by cell- or particle-bound C5 convertase (Vogt et al. 1978). C5b spontaneously associates with C6 thus stabilizing its very short-lived capacity for association with a cell membrane. The C5b-6 complex will reversibly bind to a membrane (Shin et al. 1971; Goldlust et al. 1974), and in turn spontaneously associates with C7. The resulting C5b-7 complex exposes an apolar domain towards which all three components probably contribute, and which enables its entrance and anchoring in the membrane matrix solely through hydrophobic interactions (Hammer et al. 1975; Hu et al. 1981). C5b can diffuse for a short distance from the site of generation until formation of a complex with C7 occurs. This enables the C5b-9 complex to reach nearby innocent bystander cells that can then be subject to attack. This phenomenon, termed "reactive lysis", is in fact demonstrable in vitro (Thompson and Rowe 1968; Thompson and Lachmann 1970; Lachmann and Thompson 1970), but it is not known to what extent it occurs during complement activation in a tissue. If C5b associates with C7 during transit through the tissue fluid, the ensuing C5b-7 complex will spontaneously associate with the S-protein, thereby losing its capacity for insertion in the membrane (Podack and Müller-Eberhard 1978). The association process continues, leading to formation of SC5b-9

S. Bhakdi and J. Tranum-Jensen

Table 1. Properties of native terminal complement components

	Serum concentration (μg/ml)	M_r	Electrophoretic mobility	$S_{20,w}$	Polypeptide chains
C5	150–170	200 000	β_1	8.7	2
C6	70– 80	130 000	β_2	5.5	1
C7	40– 60	120 000	β_2	6.0	1
C8	70– 80	155 000	β_1	8.0	3
C9	50– 60	70 000	α	4.5	1
S (vitronectin)	300–500	85 000	α		1

complexes.However, the exact requirements for SC5b-9 generation are not clearly defined. Somewhat surprisingly, little SC5b-9 forms as a by-product of complement activation on target erythrocytes, even though the effectiveness of the binding of C5b to cells is low (Bhakdi and Tranum-Jensen 1984a). If C5b-7 is generated and inserts into a target membrane, C8 and C9 then bind to this complex to generate the pore-forming C5b-9 complex.

5.3.3 Transmembrane Pore Formation by C5b-9

Preamble. The first studies indicating that complement generates transmembrane pores were the experiments on osmotic protection by Green et al. (1959) and Sears et al. (1964). Shortly thereafter, Borsos et al. (1964) and Humphrey and Dourmashkin (1969) discovered ring-structured "lesions" on complement-lysed membranes, and eventually concluded that these lesions corresponded to functional "hits". A major advance was made when Lachmann and Thompson (1970) identified C5-C9 as the sole complement componends involved in generating both the ultrastructural and functional lesions (Lachmann et al. 1970, 1973; Thompson and Lachmann 1970). Subsequently, Kolb and coworkers showed the formation of SC5b-9 complexes in inulin-activated serum (Kolb et al. 1972; Kolb and Müller-Eberhard 1973, 1975a), and this led Mayer (1972) to reason that each ultrastructural ring lesion represented a membrane-embedded C5b-9 complex, the hollow interior of which formed a hydrophilic transmembrane pore. This hypothesis received strong support from subsequent studies in which C5b-9 complexes were isolated from target membranes and shown to represent hollow cylindrical structures of amphiphilic nature (Bhakdi et al. 1976, 1978; Tranum-Jensen et al. 1978; Bhakdi and Tranum-Jensen 1978) which were identical to the ring lesions found by Humphrey and Dourmashkin (1969). Membrane reconstitution studies enabled the transmembrane pore to be visualized directly (Bhakdi and Tranum-Jensen 1978). Functional studies performed in the laboratories of Mayer (Michaels et al. 1976; Ramm and Mayer 1980), Lauf (Lauf 1975; Sims and Lauf 1978, 1980). Dalmasso (Giavedoni et al. 1979; Dalmasso and Benson 1981), and Borsos (Boyle and Borsos 1979; Boyle et al. 1979, 1981) yielded data consistent with the pore concept, although arguments arose about the true functional size and molecular composition of the lesions, these are still the matter of some debate. However, consensus is apparently being reached that the prime reason for the apparently conflicting observations on the functional pore size of the C5b-9 lesion derives from a genuine heterogeneity of C5b-9 complexes (Bhakdi and Tranum-Jensen 1984a; Amiguet et al. 1985; Ramm et al. 1985).

Generation of the C5b-9 Pore. The binding of one molecule of C8 to
C5b-7 occurs via the β-chain of the C8 molecule (Monahan and Sodetz
1980; Monahan and Sodetz 1981), and is a temperature-independent
process (Bhakdi and Tranum-Jensen 1986). Functional lesions (pores)
appear at this stage of assembly when a sufficiently high density of
C5b-8 complexes is generated on a membrane (Stolfi 1968; Ramm et al.
1982b). The structure and precise nature of these C5b-8 pores is unknown;
they develop with a marked time lag relative to C5b-9 lesions (Ramm et
al. 1982b), and probably ensue from secondary aggregation that is very
pronounced in the absence of C9 (Bhakdi and Tranum-Jensen 1984a;
Cheng et al. 1985). It is likely that the first C9 molecule binds to a site
on C8 αγ since C9 will not bind to C5b-7 complexes carrying C8 β
alone. At 0° C a limited association of one C9 molecule with each
C5b-8 complex takes place. At low densities these lesions do not generate
pores if kept at low temperature. Upon elevation of temperature to
37° C, C5b-9 complexes containing an average of only one C9 molecule
generate transmembrane pores that have been estimated to be 2–3 nm
in diameter by osmotic protection experiments (Bhakdi and Tranum-
Jensen 1986). Evidence has been obtained that raising the temperature
induces a conformational change in the bound C9 molecule (Boyle et al.
1978). The stable C5b-9 pores that harbor an average of one C9 molecule
per complex are entirely devoid of a ring-shaped ultrastructure (Bhakdi
and Tranum-Jensen 1986).

When membranes bearing $C5b-8_1 C9_1$ complexes are offered a surplus
of C9 at 37° C, binding of further C9 molecules occurs. A heterogeneous
array of C5b-9 complexes then forms, represented mainly by C5b-8
monomers (Tschopp et al. 1982) carrying a widely varying number
(2–16?) of C9 molecules (Bhakdi and Tranum-Jensen 1984a; Amiguet
et al. 1985; Tschopp et al. 1985). The mean ratio of C9:C8 on sheep
erythrocyte membranes approaches 5–6:1 (Kolb and Müller-Eberhard
1974; Sims 1983; Stewart et al. 1984; Bhakdi and Tranum-Jensen
1984a, 1986; Amiguet et al. 1985) and may rise to 8–9:1 on rabbit
membranes (Bhakdi and Tranum-Jensen 1984a). Concomitant to this
temperature-dependent C9 oligomerization (Sims and Wiedmer 1984),
widening of the pores occurs and ring-shaped ultrastructural lesions
appear (Bhakdi and Tranum-Jensen 1986 (Fig. 15). The increase in
functional pore size (up to 7–10 nm in diameter) is probably due to the
generation of circular lesions containing many C9 molecules. By inference,
lesions of intermediate size will probably be formed when lower numbers
of C9 molecules bind to a C5b-8 complex.

There appear to be two main causes for the appearance of heterogeneous C5b-9 lesions (Bhakdi and Tranum-Jensen 1984a). First, a

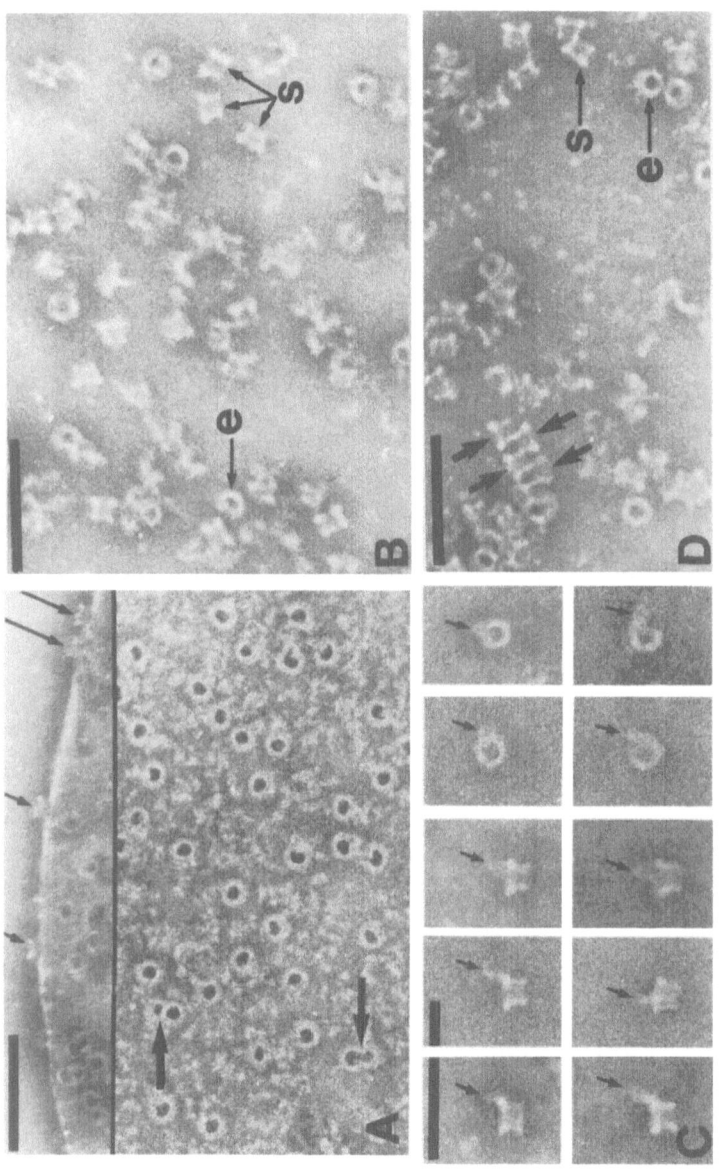

Fig. 15. A Negatively stained, complement-lysed erythrocytes. C5b-9 complexes are seen as numerous circular lesions over the membrane together with some twinned forms (*large arrows*). The complexes are seen as 10 nm high cylindrical projections along the bent edge of the ghost membrane at the top (*arrows*). The light rim representing the sharply bent membrane in tangential view is attenuated or interrupted at the site of attachment of the complexes. **B** Negatively stained preparation of isolated C5b-9 complexes in detergent solution. The complex has the basic structure of a 15 nm high, thin-walled cylinder, rimmed by an annulus at one end. The cylinder is seen in various levels of tilt between side views (*s*) and axial projection (*e*). **C** Selection of C5b-9 complexes exhibiting a small appendage (*arrows*), often seen on the annulus, particularly by low electron-dose image recording. This stalk carries antigenic determinants of C5 and C6 (Tschopp et al. 1982b). **D** "Poly-C9" formed by prolonged incubation of purified human C9 in detergent free buffer solution at 37°C as described by Tschopp et al. (1982a). The C9-oligomers exhibit a cylindrical structure closely resembling the C5b-9 complex except for the absence of appendages on the annulus. Occasionally, small ordered arrays of cylinders are seen (*arrows*), associated at the putative apolar terminus opposite the annulus. (From Bhakdi and Tranum-Jensen 1984a)

natural shortage of C9 prevails in serum, the normal ratio of C9:C8 being only 2–3:1. Hence, when the serum dose is low and C8 consumption is high, not enough C9 molecules are available to saturate out membrane-bound C5b-8 complexes. Raising the serum dose partly changes the situation since the number of C5b-8 binding sites on a membrane is necessarily limited, and a C9 surplus ensues as C8 consumption decreases. However, the maximum ratio of C9:C8 on erythrocyte membranes never approaches the theoretical maximum of 12–16:1. This may derive from the fact that the C9-C9 oligomerization process is sensitive to interruptions and, one halted, cannot be restarted. It is also possible that membrane surface constituents often sterically hinder the formation of ring-shaped oligomers. As a consequence, only a minority of complexes ultimately contain the high number of C9 molecules required to generate a fully circular lesion. Most complexes will exhibit either defects in the ring structure, or no recognizable ring structure at all.

Role of Poly-C9. The term "poly C9" has been used to designate SDS-resistent, high molecular weight C9 complexes that show a ring ultrastructure in the electron microscope (Podack and Tschopp 1982a, b; Tschopp et al. 1982a). Poly-C9 was originally prepared by incubation of purified C9 at 37°C for prolonged time; it is now recognized that C9 oligomerization is dependent on the presence of divalent cations, in particular, Zn^{2+} (Tschopp 1984). The ring ultrastructure of poly-C9 is virtually identical to that of C5b-9 membrane lesions, which are known to contain multiple C9 molecules. Further, poly-C9 can attach to liposomes and form pores in these artificial membranes (but not in cells). Hence, poly-C9 was thought to carry the primary membrane-damaging function within the C5b-9 complex (Podack et al. 1982; Tschopp et al. 1982a; Müller-Eberhard 1984). The molecular weight of poly-C9 formed from purified C9 in vitro is $10^6 \pm 1.5 \times 10^5$ and hence a poly-C9 tubule harbors 12–18 molecules of C9 (Tschopp et al. 1984).

It should be noted that only a small portion of membrane-bound C5b-9 complexes forming on erythrocytes carry C9 in a SDS-resistant, high molecular weight form comparable with that of poly-C9 (Yamamoto et al. 1982; Bhakdi and Tranum-Jensen 1984a); this is in accord with the relatively low average ratio of C9:C8 found on biological target membranes (Sims 1983; Stewart et al. 1984; Bhakdi and Tranum-Jensen 1984a, 1986; Amiguet et al. 1985). Possibly, the development of resistance to dissociation by SDS of the oligomeric structure within C5b-9 is time dependent and develops secondary to pore formation. Indeed, there is disagreement about the composition of the SDS-resistant moiety within C5b-9 complexes. Thus, although initial studies referred to this moiety as poly-C9, Podack (1984) more recently reported that other terminal components (C5-C9) were also present in the aggregate.

Therefore, it now seems that the original proposal that all five terminal complement proteins actually participate in forming, the cylindrical, membrane-embedded portion of the C5b-9 complex is being revived (Bhakdi and Tranum-Jensen 1979; Bhakdi et al. 1980). In fact, there are many lines of evidence to support this view. First, apolar photolabeling clearly indicates that parts of C5-C8 polypeptide chains become inserted in the membrane up to the stage of C5b-8 assembly. Upon addition of C9 to the complex, the latter predominantly picks up the labels. However, as pointed out by Amiguet et al. (1985), the extent of labeling of other terminal components, as also found by Hu et al. (1981) and Steckel et al. (1983), is significant. Secondly, although the C5b-8 complex was initially localized by immuno-electron microscopy mainly to an extramembraneous stalk structure attached to poly-C9 (Tschopp et al. 1982b), it is apparent that the volume of this structure is far too small to account for the entire mass of C5b-8. Finally, the fact that C5b-9 complexes devoid of poly-C9 and containing very low numbers of C9 molecules can still generate stable, rather large transmembrane pores constitutes indirect evidence that C5-C8 components must also contribute towards formation of the pores.

To conclude, it is clear that structures akin to poly-C9 pores form when target cells are attacked by C5b-9. However, it is not established whether the composition of these is identical to that of poly-C9, or whether they harbor additional terminal complement components. It is certainly an oversimplification to state that the membrane-damaging function of C5b-9 is carried by poly-C9. The majority of lesions represent heterogenous functional pores, although they do not contain the high molecular weight, SDS-resistant, and poly-C9-related structures.

Depth of Bilayers Penetration by C5b-9. Electron-microscopical measurements of the height of fully formed C5b-9 cylinders in isolation compared with the height projection from the lipid bilayer (Fig. 5, 15) gave an early indication that 4–5 nm of the height of the cylinder was available for burial within the bilayer (Tranum-Jensen et al. 1978; Bhakdi and Tranum-Jensen 1978). Freeze-etch electron microscopy corroborated this (Tranum-Jensen and Bhakdi 1983), and analysis of complementary fractures showed that the intramembraneous portion of the cylindrical C5b-9 complexes extended into the lipid plateau of fracture PF faces (Fig. 6). Moreover, comparison of etched and unetched fractures confirmed the existence of water-filled pores traversing the interior of the C5b-9 cylinders.

Subsequently, evidence for the penetration of C9 to the cytoplasmic face was found by Morgan et al. (1984) and Whitlow et al. (1985). The latter used protein cross-linking at the cytoplasmic face to demonstrate the transmembrane orientation of C9 in C5b-9 complexes. Morgan et al. (1984) reported that a monoclonal antibody reacted with a C9 epitope that was exposed only at the cytoplasmic membrane face. It now seems important to have these studies fully confirmed, because epitope mapping locates the portion of C9 that reacts with the monoclonal to a region that in all probability is situated outside the membrane (Stanley et al. 1986). A two-step model for C9 insertion has been proposed to account for this serious discrepancy (Stanley et al. 1986), but in our view the model is not likely to be correct.

5.3.4 Past and Present Controversies about the Complement Lesion

A confusing array of controversies has surrounded the issue of complement-mediated membrane damage and a brief synopsis is warranted at this stage. The first controversy was related to the question of whether membrane damage was caused by formation of a lipid-lined pore, induced through lipid disorganization around inserted C5b-9 complexes, or whether the C5b-9 complex itself forms the pore. This controversy is essentially settled today, and it is generally agreed that the protein pore concept is correct (Mayer et al. 1981; Bhakdi and Tranum-Jensen 1983; Müller-Eberhard 1984).

The second controversy was related to the composition of C5b-9 lesions. A view that each C5b-9 pore consisted of C5b-8 dimers plus bound C9 oligomers drew attention at one time, but was not supported in subsequent studies. Today, there is increasing evidence that the bulk of C5b-9 complexes contain one molecule of C5b-9 (Bhakdi and Tranum-Jensen 1981; Ramm et al. 1982a; Tschopp et al. 1982b; Silversmith and Nelsestuen 1986), and the dimer concept has been abandoned (Müller-Eberhard 1984).

Another controversial issue was related to the cause and nature of the C9-dependent heterogeneity of the complement lesions. Podack et al. (1982) proposed that lesion heterogeneity was due to the parallel presence of monomeric and dimeric terminal complexes on target membranes. As the dimer model has been abandoned, this concept is also untenable. Esser (1981) and Sims and Lauf (1980) have proposed that aggregation of individual C5b-9 lesions into heterogeneous supramolecular aggregates is primarily responsible for lesion heterogeneity. We and Amiguet et al. (1985) favor the view that heterogeneity is due to the presence of C5b-9 complexes that contain varying numbers of C9 molecules. This concept appears compatible with functional data obtained by Boyle et al. (1979, 1981) and Ramm et al. (1982a, 1985). Aggregation of C5b-9 might contribute to lesion heterogeneity, but it is our view that this effect is not of prime importance and that the gross heterogeneity in molecular mass of detergent-solubilized C5b-9 (Bhakdi and Tranum-Jensen 1981, 1984a; Ware et al. 1981) is mainly a product of secondary aggregation.

5.3.5 On the Effectiveness of C5b-9 Pore Formation in Erythrocyte Membranes

An issue that is important to the understanding of many immunopathological phenomena relates to the effectiveness of C5b-9 pore formation. Will each C5b-9 complex generated on an erythrocyte membrane form a transmembrane pore, or are "abortive" attachments possible? This question is of interest because there are numerous examples where C5b-9 complexes appear to differ in their hemolytic efficiency; a given total of terminal complexes will cause extensive hemolysis in one instance and little in another. For example, C5b-9 deposition invoked by complement-fixing autoantibodies (e.g., cold agglutinins) appears to be of low efficiency, whereas C5b-9 attack induced by drug-dependent antibodies is comparatively more efficient (Salama et al. 1987). Another example is the relatively high susceptibility of certain pathologic cells to C5b-9 attack, as exemplified by PNH erythrocytes (Rosse and Dacie 1966). In this case, any mode of complement activation will lead to enhanced lysis of a given PNH cell population compared with normal cells (Rosse and Dacie 1966; Rosse et al. 1974). Finally, homologous cells appear to be less susceptible to lysis by C5b-9 than heterologous cells (Hänsch et al. 1981).

At this stage, it is necessary to differentiate between two different mechanisms that could be responsible for the above observations. First, individual C5b-9 complexes may differ in their lytic capacity, i.e., some complexes may fail to insert into membranes and thus be nonlytic. In this case, it is conceivable that certain complement activators might induce the generation of a larger portion of lytic complexes than others (e.g., drug-dependent antibodies versus cold agglutinins). Furthermore, membrane factors dependent on membrane composition could act to restrict protein insertion; hence, certain membranes (e.g., those of PNH cells) would be more susceptible to C5b-9 attack if they lacked one or more putative "regulators".

A second possibility is that the varying C5b-9 lytic efficiency derives, at least in part, from distributional differences between the complexes on cells of a given target population. Thus, input of a given total of C5b-9 may cause a high degree of hemolysis if the complexes are evenly distributed amongst the cells of a target population. Less hemolysis may occur if the complexes are deposited on only a few cells. Both humoral (e.g., antibody) and membrane regulatory factors could affect the distribution of C5b-9 complexes. Low-affinity antibodies and immune complexes may function on a "hit-and-run" basis, depositing few C5b-9 complexes on each cell, whereas high affinity antibodies (cold agglutinins) remain restricted to certain cells and cause deposition of large numbers of complexes on relatively few cells; as a result, net hemolysis is relatively small. It is conceivable that membrane factors may also be responsible for restricting the uptake of terminal complexes, and lack or loss of these hypothetical factors (PNH erythrocytes?) might be partially responsible for their increased susceptibility to reactive lysis with C5b-9.

At present, there are data to support both mechanisms of varying C5b-9 efficiency. The "distributional" model was proposed on the basis of studies with drug-dependent antibodies versus cold agglutinins (Salama et al. 1987). It was found that cold agglutinins, presumably because of their high-affinity interaction with target cells, caused a very skewed distribution of C5b-9 within a given target-cell population; for a given total input of C5b-9 relatively few cells were lysed, but these cells carried a disproportionately high number of C5b-9 complexes. In contrast, drug-dependent antibodies apparently deposited relatively low numbers of complexes on a large number of target cells, thus causing much lysis with a low total C5b-9 input. Hence, higher lytic efficiency of drug-dependent antibodies probably does not stem from a true variation in lytic efficiency of the C5b-9 complexes per se, but from distributional differences of these complexes on the cells. Very skewed distributions of the number of lesions visible by electron microscopy were indeed observed in the early studies of Humphrey and Dourmashkin (1969) on sheep erythrocytes coated with low doses of antibody. Similarly skewed distributions were also found on rabbit erythrocytes lysed with human serum (Bhakdi and Tranum-Jensen 1984a). In the latter case, the uneven distributions could have been due to membrane factors, such as those affecting the generation and function of the alternative pathway convertase.

Evidence that C5b-9 complexes themselves may differ in their pore-forming capacity comes from studies on PNH cells, and from studies of homologous species restriction of C8- or C9-dependent lysis.

Studies on PNH Erythrocytes. Four types of PNH have been defined according to the pattern of afflicted erythrocytes found in the circulation (Rosse et al. 1974; Logue et al. 1974). Of these, type III cells have been studied in most detail with respect to complement-induced lysis. The early work of Rosse et al. (1966, 1974) established that these cells bound more C3 and were basically more susceptible to lysis by complement. Recently, it was found that these cells have a defect in an early regulatory factor (DAF) which would explain the increased C3 binding (Nicholson-Weller et al.1983; Pangburn et al. 1983). However, this defect alone could not fully explain the PNH anomaly because cells were also highly susceptible to reactive lysis in the absence of C3 (Packman et al. 1979; Rosenfeld et al. 1985); hence, PNH cells were apparently defective or deficient in one or several unknown regulatory factor(s) operating at a stage after C3. One study has indicated C5b binding may be regulated by the CR-1 (C3b) receptor (Fisher et al. 1984), but this appears not to be defective in PNH cells (Roberts et al. 1985). At present, it is not clear whether PNH cells in fact bind more C5b-9, or whether C5b uptake is comparable to that of normal cells but the deposited C5b-9 complexes are hemolytically more efficient on PNH erythrocytes. In fact, somewhat most recent studies have yielded conflicting results. Hu and Nicholson-Weller (1985) and Rosenfeld et al. (1985) reported comparable binding of C5b-9 to PNH versus control erythrocytes, but more C9 "insertion" (as determined by an apolar photolabel) into the former cells (Hu and Nicholson-Weller 1985). In contrast, Parker et al. (1985) reported a marked increase of C5b-9 binding to PNH cells compared with normal cells under similar conditions of reactive lysis. The search for putative membrane regulatory factors whose deficiency is responsible for the PNH phenotype is still underway.

Homologous Species Restriction. Hänsch et al. (1981) presented data showing that cells laden with C5b-7 complexes were lysed well by heterologous C8 plus C9, but poorly with homologous C8 plus C9. Indirect evidence was given that human C8 plus C9 bound to the poorly lysed homologous cells; hence, hemolytically inactive or inefficient C5b-9 complexes must have formed on these erythrocytes. These data make it probable that homologous cells are able to withstand the attack of a limited number of C5b-9 complexes. More studies in this area are necessary and a number of basic questions remain to be answered. For example, despite the extensive studies involving hemolytic titration of individual terminal complement components, hardly any studies have been undertaken to determine the number of protein molecules that are required to form a functional "hole" in erythrocytes. Sims (1983) reported that approximately 300 molecules of C5b-9 per cell were needed to bind to homologous (human) erythrocytes to create

pores that would allow transmembrane passage of sucrose. Apart from
this study, no other data are available. In particular, no data are available
on the number of C5b-9 complexes required to induce hemolysis of
heterologous or homologous erythrocytes. It also appears necessary
to isolate and characterize the composition of nonlytic complexes
(e.g., their C9:C8 ratio), and, if possible their structures. Finally, membrane regulatory factors responsible for the homologous species retric-
tion need to be identified. Some recent reports indicate that such work
is in progress (Schönermark et al. 1986), although improved characterization of the putative regulatory factor(s) at a molecular level is definitely
called for.

5.3.6 Fluid-Phase (Plasma) Regulators
of the Terminal Complement Pathway

Although the major thrust of investigations on regulation of the terminal
complement pathway is presently at the level of regulatory membrane
factors (e.g., Schönermark et al. 1986; Shin et al. 1986), the possible
participation of plasma factors in protecting cells from C5b-9 attack
still requires attention. At present, the S-protein (vitronectin, see p 196)
is thought to be the major inhibitory plasma protein exerting a C5b-9
inhibitory function (Podack and Müller-Eberhard 1978; Dahlback and
Podack 1985). Before the discovery and isolation of the S-protein,
McLeod and coworkers published a series of papers on a C567 inhibitor
present in human serum (McLeod et al. 1975a, b). These studies resulted
from the initial observation of Thompson and Lachmann (1970) of the
presence of (a) plasma inhibitor(s) of reactive lysis. The C567 inhibitor
eluted in several peaks with ion-exchange chromatography. In contrast,
the S-protein behaves as a more homogenous, acid pseudoglobulin,
eluting in one broad peak upon DEAE chromatography. Following the
isolation and characterization of the S-protein, no further reports have
appeared on the C567 inhibitor and the tacit assumption made by many
researchers is that the C567 inhibitor originally described by McLeod
et al. (1975a, b) was represented in the main by the S-protein. However,
the physicochemical properties of the C567 inhibitor certainly do not
fully conform to those of the S-protein, and the possibility remains that
inhibition of C5b-9 may occur via the action of additional plasma
components. In this connection, it is noteworthy that Sassi et al. (1987)
have reported the presence of a potent C5b-9 inhibitor in serum of
Ctenodactylus gondi that inhibits binding of C5b-9 to cells in a manner
distinct from that of the S-protein. It will be of interest to determine
whether a similar anticytolytic factor is also present in sera of other

animal species, and whether it may be related to a component of the C567 inhibitor alluded to earlier.

5.3.7 Attack on Nucleated Cells by C5b-9

In view of the rapid turnover of membrane in nucleated cells, it is not surprising that these cell are capable of "repairing" a limited number of complement lesions. Early indications of this were obtained by Green et al. (1959), who reported the release of low molecular weight markers from complement-attacked nucleated cells without overt cell lysis. Ohanian, Schlager, Boyle, and Borsos later conducted a number of studies collectively showing the dependence of cell recovery on a number of factors, including lipid turnover and cAMP generation (for review see Ohanian and Schlager 1981). These studies did not include a biochemical characterization or quantitation of the C5b-9 complexes assumed to be forming on the cell targets. Data clearly showing a requirement for multiple "hits" for induction of lysis of nucleated cells were presented by Mayer's laboratory (Koski et al. 1983; Ramm et al. 1983). These studies demonstrated that nucleated cells were capable of tolerating attack by limited numbers of C5b-9 complexes, and that subcytolytic numbers of terminal complexes could be removed from the surface the the cells. Carney et al. (1985) showed that one mechanism for removal of C5b-9 complexes from heterologous cell surfaces probably involved endocytic uptake. Independently, Campbell and Morgan (1985) also reported removal of complement complexes from cell surfaces, whereby evidence for "shedding" of afflicted membrane areas was presented. There is evidence to suggest that membrane repair is dependent on influx of Ca^{2+} (probably through the transient complement pore), since removal of extracellular Ca^{2+} reportedly decreased the repair capacity (Morgan and Campbell 1985; Carney et al. 1986).

Taken together, there is currently no doubt that nucleated cells can repair complement lesions. There are several interesting points that merit further study, e.g., the contribution of exo-versus endocytic processes in membrane repair, the fate of both exocytosed and endocytic vesicles, and the possible significance of endolysosomal leakiness that may ensue from fusion of lysosomes with pore-containing endosomes. The examination of C5b-9 action on autologous rather than heterologous cells is also desirable since this type of situation probably arises quite often in human disease.

5.3.8 Secondary Effects Elicited by Subcytolytic Doses of C5b-9 in Nucleated Cells

An important aspect relates to secondary effects on nucleated cells evoked by subcytolytic complement doses. First reports in this area were made by Imagawa et al. (1983, 1986), who showed that terminal complement components elicited activation of the arachidonic acid cascade. Similar data were obtained by Hänsch et al. (1984, 1985) with macrophages and platelets. Activation of the arachidonic acid cascade by C5b-9 complexes appeared phenomenologically similar to that elicited by staphylococcal α-toxin (Suttorp et al. 1985a). Subsequent studies have shown that the activation step is probably dependent on Ca^{2+}-influx, presumably via C5b-8/9 pores (Seeger et al. 1986). A clear dissociation between arachidonic acid cascade activation and cytolytic C5b-9 effects in homologous polymorphonuclear leukocytes could be demonstrated (Seeger et al. 1986). In addition, these studies demonstrated a dissociation between C5a effects and the release of thromboxane, which was dependent on pore formation. The biological significance of these findings lies in their probable contribution towards induction of pathophysiological changes during any disease involving auto-attack of tissues by complement. In another recent report, C5b-9 was shown to stimulate production by cultured rat mesangial cells of reactive oxygen metabolites (Adler et al. 1986).

5.3.9 C5b-9 and Disease

Novel antigenic determinants (neoantigens) become exposed on terminal complement components when they assemble into fluid-phase or membrane C5b-9 complexes (Kolb and Müller-Eberhard 1975; Bhakdi et al. 1978, 1983b). The majority of neoantigens are carried on C9 molecules. Poly- and monoclonal antibodies directed against a neoantigen permit the unambiguous identification of terminal complexes both in plasma and on cell membranes and tissues. Use of such antibodies has led to the detection of terminal complement complexes in a wide variety of tissues (Table 2), and the possibility is being considered that C5b-9 attack on autologous membranes may play a causative role in the pathogenesis of various immunological disorders (for example see Biesecker 1983; Couser et al. 1985). This is likely to be correct in many cases. The first supportive evidence was found by Salant et al. (1980), who showed that C3 depletion in rats by administration of cobra venom factor completely prevented the development of proteinuria in passive Heymann nephritis without altering antibody deposition in the glomeruli. Subsequently, a number of studies of experimental

Table 2. Immunohistological detection of C5b-9 neoantigens in tissues

Organ/tissue	Reference	Disease/experimental model
Nervous system	Sahasi et al. (1980)	Myasthenia gravis
Kidney		
Human renal disease	Biesecker et al. (1981)	SLE
	Falk et al. (1983)	SLE, diabetes, amyloid
	Parra et al. (1984)	Post-streptococcal glomerulonephritis
	Hinglais et al. (1986)	Normal kidneys and diverse classified renal diseases
	Cosyns et al. (1986)	Transplant reject
Experimental models	Koffler et al. (1983)	Chronic serum sickness
	Adler et al. (1984)	Passive Hyman nephritis
	Biesecker et al. (1984)	Active Hymann nephritis
	Perkinson et al. (1985)	Passive Hymann nephritis
	De Heer et al. (1985)	Active Hymann nephritis
Striated muscle	Engel and Biesecker (1982)	Muscle fiber necrosis
Lung	Kopp and Burrell (1982)	Antibody-coated alveolar basement membrane
Skin	Biesecker et al. (1982b)	SLE
	Dahl et al. (1984)	Bullous pemphigoid
Vessels	Vlaicu et al. (1985)	Arteriosclersosis
	Hinglais et al. (1986)	Normal arteries
	Rus et al. (1986)	Arteriosclerosis
	Kissel et al. (1986)	Dermatomyositis
Heart	Schäfer et al. (1986)	Myocardial infarction

models have corroborated the idea that the terminal complement pathway is involved in invoking proteinuria in several models of immunologically mediated renal injury (for review see Couser et al. 1985). In addition to the direct demonstration of C5b-9 neoantigens in capillary walls of the afflicted organs, it has also been shown that the extent of proteinuria is diminished in animals deficient in a terminal complement component (Groggel et al. 1983, 1985; Cybulsky et al. 1986).

Clearly, attack of cells by autologous C5b-9, elicited for example by autoantibodies, must be generally detrimental, and secondary processes such as release of arachidonate metabolites would aggravate inflammatory reaction. Nevertheless, the mere presence of C5b-9 in tissues should not be equated with primary C5b-9-mediated cell membrane damage for several reasons. First C5b-9 deposition may occur as a secondary reaction, e.g., on dead or dying cells because of loss of regulatory anticomplement membrane factors. It is noteworthy that C5b-9 has been detected on necrotic muscle fibers (Engel and Bisecker 1982), and in infarcted areas of human myocardium (Schäfer et al. 1986). Secondly, C5b-9 deposition, e.g., in the basement membrane of glomerula, may not create pores at all since the attack does not occur on bilayer targets. Whether and how tissue damage can occur in such cases still requires clarification. Finally, no study to date has formally identified tissue-localized C5b-9 as the membrane complex, since all neoantigen-specific antibodies used also react with SC5b-9 complexes (see Bhakdi and Muhly 1983; Bhakdi et al. 1983b; Hugo et al. 1985). Hence, the additional application of antibodies to the complement S-protein (vitronectin) will be required to differentiate C5b-9 membrane complexes from SC5b-9. These analyses should be carried out because recent data have disclosed an unexpected connection between the function of the S-protein in the complement and coagulation cascades (Jenne et al. 1985; Preissner et al. 1985; Podack et al. 1986), and its role as a cell adhesion factor. Studies on the S-protein at a molecular genetical level have shown it to be identical with vitronectin, the "serum-spreading factor" (Jenne and Stanley 1985). Vitronectin belongs to the family of substrate adhesion molecules which include fibronectin, laminin, and chondronectin (Holmes 1967; Barnes et al. 1983; Hayman et al. 1982, 1983; Barnes and Silnutzer 1983). It is known that the protein possesses both cellular binding sites and sites that bind to heparin-like molecules. The S-protein could therefore serve to "target" terminal C5b-9 complexes to specific cell and tissue sites. Further, it has been found that the protein may act in a pro-coagulatory fashion by protecting thrombin from the inactivating effects of antithrombin III (Jenne et al. 1985). Overall, these findings lead to new concepts regarding the biological functions and interactions of the terminal complement and coagulation cascades. In particular, they indicate that both cascades may be functionally interrelated, particularly at sitzes of endothelial lesions where there is activation of the coagulation system and egress of complement components into the connective matrices. It is of interest to note that neoantigens of C5b-9 have been repeatedly detected in such connective matrices within arterial walls in the absence of recognizable pathophysiological alterations (Hinglais et al. 1986; Schäfer et al. 1986), and monoclonal antibodies to vitronectin have led to the detection of this molecule in similar locations (Hayman et al. 1983). Hence, it is possible that terminal complement complexes are rapidly cleared from plasma, partially by deposition in connective tissues and on cell surfaces bearing receptors for vitronectin. The function of such deposited complexes remains unknown. It is noteworthy that SC5b-9 exhibits the same capacity to induce cell-spreading factor as native S-protein (K. Preissner and S. Bhakdi, unpublished work), and this property may be biologically relevant at sites of SC5b-9 deposition in tissues.

Recent data indicate that the membrane C5b-9 complex is selectively deposited in large amounts in infarcted areas of human myocardium (Fig. 16; Schäfer et al. 1986). This finding corroborates earlier data that indicated a role for complement in the pathogenesis of secondary, heterolytic damage occurring during myocardial infarction (e.g., Hill and Ward 1971; Maroko et al. 1978; Pinckard et al. 1980; McManus et al. 1983). In these studies, S-protein-specific monoclonal antibodies yielded only very weak staining, indicating that the detected neoantigens derived from the presence of the membrane C5b-9 complex. In a few instances staining was selectively localized to the plasma membrane of the ischemic myocardial cells. It was therefore possible that deposition of C5b-9 represented a truly causal factor invoking heterolytic myocardial damage during the infarction process. An intriguing aspect emerging from these findings is that complement attack and C5b-9 deposition on autologous cells may occur via nonimmune mechanisms, e.g., on dead or dying cells in general, probably due to failure of regulatory factors (see for example Fearon 1985) to prevent spontaneous autoattack. It is even possible that complement components, including C5b-9, thus fulfil a "scavanger" function within the organism and aid in the removal of dead or dying cells.

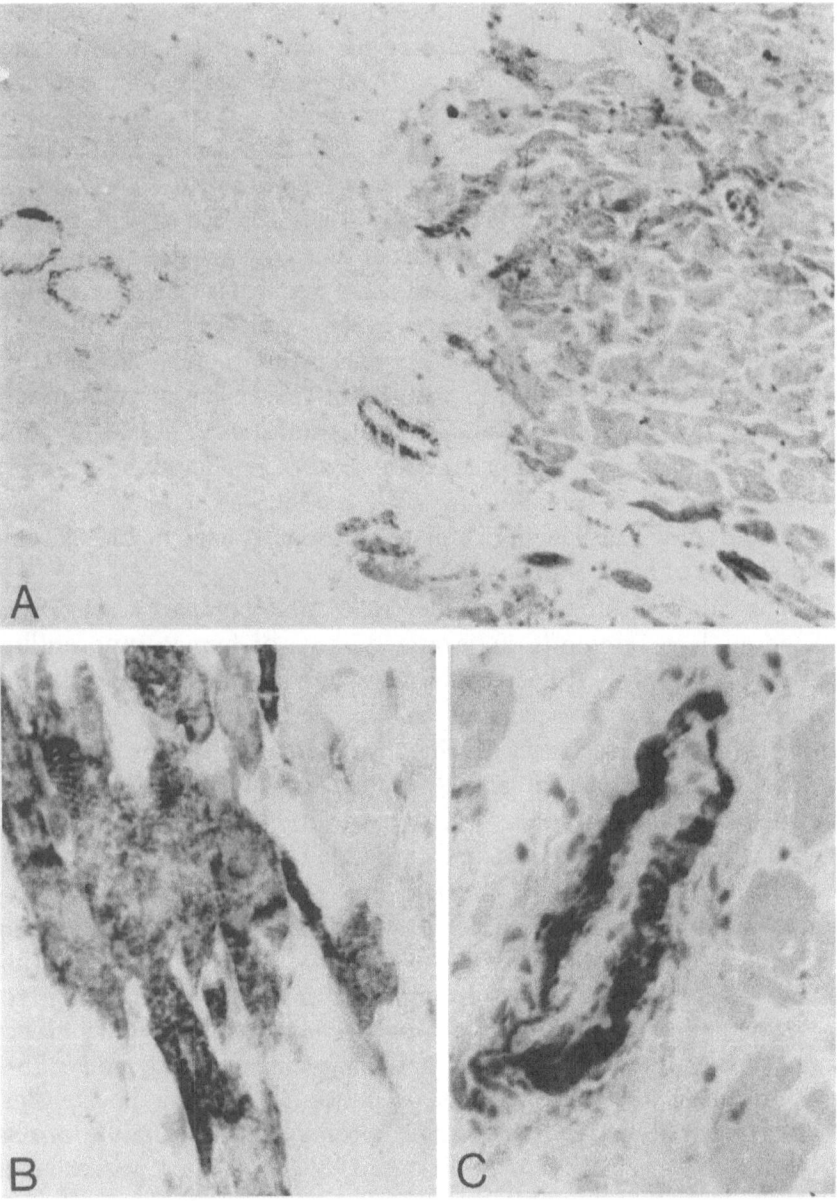

Fig. 16 A–C. Immunocytochemical demonstration of C5b-9 complement complex in myocardial infarction. Frozen sections of autopsy material; polyclonal antibody to C5b-9 neoantigens. Estimated age of infarction about 10 h, as concluded from early start of granulocytic infiltration. Distinct staining of arterial walls and of areas with myocardial infarction. Negative reaction in normal myocardium. **A** Clear demonstration of infarcted areas even at low magnification (✕ 140). **B** Intense cytoplasmic staining of whole necrotic cells (✕ 345). **C** Staining of subendothelial layers in artery walls also in non-infarcted regions (✕ 345) (From Schäfer et al. 1986)

5.3.10 Hereditary Deficiencies in Terminal Complement Components

A total inability to generate the C5b-9 complex due to a genetic deficiency for synthesis of one of the terminal components need not have lethal effects. Deficiencies in C5 (Rosenfeld et al. 1976; Synderman et al. 1979; Haeney et al. 1980), C6 (Leddy et al. 1974; Lim et al. 1976), C7 (Boyer et al. 1975; Nemerow et al. 1978; Loirat et al. 1980), C8 (Petersen et al. 1976; Jasin 1977; Tedesco et al. 1980, 1983; Densen et al. 1983), and C9 (Lint et al. 1980) have been reported. In one type of functional C8 deficiency, the biochemical defect has been defined as the entire absence of the C8β chain (Tschopp et al. 1981). The salient clinical feature of C5-C8 deficiency is an increased susceptibility to infections by *Neisseria* (Lee et al. 1978; Haeney et al. 1980; Loirat et al. 1980; Densen et al. 1983). In addition, one report has indicated a connection between C8 deficiency and systemic lupus erythematosus-like disease (Jasin 1977). Since all the defects have been discovered only recently, prolonged further periods of observation will be required to assess the true significance of the terminal pathway in humans.

5.4 Pores Generated by Lymphocytes

5.4.1 Biological Significance

Two characterized classes of lymphocytes, namely, cytotoxic T-lymphocytes (CTLs) and large granular lymphocytes (LGLs) are endowed with the potential to kill target cells (for review see Henkart 1985). The former carry T-lymphocyte markers, exhibit MHC (major histocompatibility complex) restriction, and target recognition occurs via the T-cell receptor. Cytolysis by CTLs is the major mechanism of transplant rejection and an important mechanism for killing of virally infected tissue cells. LGLs carry some T-cell markers, but target recognition is MHC independent and occurs either via Fc receptors (i.e., antibody-dependent cellular cytotoxicity, ADCC) or via other, less well-defined receptors (e.g., natural killer (NK) cells). Both CTLs and LGLs harbor cytoplasmic granules that contain proteins which are capable of forming pores in target membranes, this probably being a major effector mechanism of the lymphocytes. The cytotoxic action of both classes of lymphocytes appears to comprise a common pattern of events involving:

1. Target recognition
2. Establishment of close cell-to-cell contact between target and effetor cell

3. Polarization of cytoplasmic components in the effector cells with localization of most cytoplasmic organelles between the nucleus and the target cell
4. A secretory process leading to exocytosis of granular contents into the intercellular, space between effector and target cell; this process is Ca^{2+} dependent and exocytosis is probably triggered by influx of Ca^{2+} into the attacking lymphocytes
5. Formation of large transmembrane pores in the target cell membrane through assembly of the granule protein protomers into circular pore structures; this process appears phenomenologically similar to the formation of C9, α-toxin, and streptolysin-O pores
6. The possible subsequent occurrence of one or several secondary phenomena, including triggering of release of DNAase, as factors potentiating the cytotoxic process

Recent reviews should be consulted for details of these events, and we will here confine ourselves to a brief synopsis of the membrane-damaging step. Henney (1973, 1974), Martz (1976), Sanderson (1976), and Sanderson and Thomas (1977) provided evidence that cytotoxic lymphocytes inflicted damage to target cell membranes, resulting in the release of small molecules such as [86]Rb^+ and ATP. Henkart and Blumenthal (1975) introduced the pore concept of membrane damage when they demonstrated large, discrete increases in ionic permeability in artificial planar bilayers induced through attack in the ADCC system. Evidence for pore formation by lymphocytes in a biological cell membrane was subsequently obtained by Simone and Henkart (1980). These investigators used resealed erythrocyte ghosts as ADCC targets and were able to demonstrate the generation of large transmembrane pores which appeared from molecular sieving experiments to be approximately 15 nm in diameter. At the same time, Dourmashkin et al. (1980) published the first description of circular "lesions" in the target membranes using the same experimental system. These ring structures had an internal diameter of 15 nm and were reminiscent of complement lesions. Podack and Dennert (1983) and Dennert and Podack (1983) later confirmed the presence of ring structures on erythrocyte target membranes lysed by cloned NK cells. These investigators detected ring structures of two sizes and showed these to be stable after extraction in detergent solution.

Recently, Henkart et al. (1984) and Millard et al. (1984) reported the purification of cytoplasmic granules from rat LGL tumors and demonstrated their lytic effect on a variety of cells. Granules isolated from noncytolytic lymphoid cells did not have the cytotoxic activity. The same investigators found that purified granules would perturb liposomal membranes containing no protein, causing release of trapped carboxyfluorescein (Blumenthal et al. 1984). Concomitantly, the typical 15 nm cylindrical structures were found attached to the liposomal bilayer. Antibodies raised against granular proteins were found to inhibit the cytotoxic action of the LGL (P. Henkart, personal communication). Masson and Tschopp (1985) have recently isolated and identified a protein of M_r 66 000 from cytolytic T-cell granules as the protomer of the pores. The protein exhibited several physicochemical properties similar to C9. Insertion into lipid of the membrane-bound protein was demonstrated by labeling with a membrane-restricted photoactivable probe.

Experiments describing the isolation and functional characterization of the lymphocyte-pore-forming protein have subsequently also been described by Podack and Königsberg (1984), Podack et al. (1985) and Young et al. (1986 a–d). Collectively, the data speak for a true causal relationship between the presence of pore-forming proteins in the lymphocyte granules, and lymphocyte-mediated cellular cytotoxicity. The reader is referred to a recent review (Henkart 1985) on the granule exocytosis model. Future work should soon lead to the detailed characterization of the granular proteins that are involved in pore formation. It is conceivable that these proteins are molecularly related or similar to terminal complement components (in particular C9). Indeed, preliminary data to this effect have appeared (Ward and Lachmann 1985; Young et al. 1986c; Zalman et al. 1986). Issues to be resolved relate to the question of specific binder molecules for the lymphocyte pore formers. It is also an enigma why the proteins attack the target cells and not the effector cells themselves, since the latter are equally susceptible to lysis by the granular proteins. Finally, the possibility that secondary events such as release of DNAse and/or proteases from the effector cells play important supportive roles in the cytolytic process warrants further study.

6 Pores with No Recognizable Ultrastructure

6.1 Cytolysins of Gram-Negative Bacteria

6.1.1 E. coli Hemolysin

Biological Significance. The hemolysin of E. coli has attracted much attention in the past, for two main reasons. First, it appears to represent a significant pathogenetic factor of the bacteria; the majority of E. coli strains causing pyelonephritis and septicemia in the newborn produce hemolysins, and toxin-producing strains also exhibit greater virulence than non-producers in animal-model experiments (for example see Fried et al. 1971; Welch et al. 1981; van den Bosch et al. 1982a, b; Waalwijk et al. 1983; Hacker et al. 1983; Hughes et al. 1983; Cavalieri et al. 1984; Welch and Falkow 1984). Secondly, the toxin represents the only known protein that appears to be genuinely secreted by *E. coli*. Hence, it is an interesting model with which to study protein transport across the inner and outer membranes of gram-negative bacteria (Springer and Goebel 1980; Goebel and Hedgepeth 1982; Wagner et al. 1983; Mackman et al. 1986).

Properties of Native Toxin. Despite the fairly extensive studies undertaken on the structure of *E. coli* hemolysin at a molecular genetic level (e.g. Goebel and Hedgepeth 1982; Wagner et al. 1983; Mackman and Holland 1984; Felmlee et al. 1985; Nicaud et al. 1985; Mackman et al. 1986), surprisingly little is known on the properties of the native protein or the mechanism of cell attack. Secretion of active toxin into bacterial culture supernatants is only detectable with appropriately selected strains. Studies performed with these strains have identified native toxin as a single-chained polypeptide of M_r 107 000–110 000 (Mackman and Holland 1984; Gonzalez-Carrero et al. 1985; Felmlee et al. 1985). Spontaneous inactivation of the toxin occurs, for unknown reasons, rapidly at 37°C and more slowly at lower temperatures. Further, ageing of toxin preparations causes degradation to occur, with the appearance of a major, non-hemolytic product of $M_r \approx 60\,000$.

Pore Formation by E. coli Hemolysin. After binding to target membranes, *E. coli* hemolysin becomes refractory to elution with 1 mM EDTA and hence is probably inserted into the lipid bilayer. Toxin-attacked cells can be protected from osmotic lysis by extracellular dextran 4, but not by sucrose or raffinose. There is rapid efflux of K^+, and influx of Ca^{2+}, mannitol, and sucrose, into dextran-protected cells (Fig. 7). These data indicate the presence of transmembrane pores of 1.5–3.0 nm diameter generated by *E. coli* hemolysin in the erythrocyte membrane (Bhakdi et al. 1986).

Sucrose density gradient centrifugation of DOC-solubilized target erythrocyte membranes results in the recovery of detergent-solubilized toxin, exclusively in monomer form (Fig. 3). Electron-microscopic examination of toxin-treated membranes has revealed no alteration compared with controls. Hence, pores formed by the toxin are tentatively proposed to represent toxin monomers inserted into the membranes; however, the possibility cannot be excluded that toxin oligomers are formed in the membrane that are dissociated by DOC (Bhakdi et al. 1986). Although the primary structure of the toxin is known (Felmlee et al. 1985), the membrane-inserted regions of the molecule have yet to be identified.

Recent studies with planar lipid membranes have shown that *E. coli* hemolysin generates transmembrane pores of 2–3 nm diameter in bilayers of pure phosphatidylcholine, but that pore formation is dependent on the presence of a membrane potential. The *E. coli* hemolysin is the first pore-forming bacterial cytolysin acting on mammalian cells that has been found to exhibit this property (G. Menestrina et al., unpublished work). In this regard, it displays similarities to pore-forming colicins (Schein et al. 1978; Konisky 1982; Cramer et al. 1983). At the same time, the above results indicate that the toxin does not require a specific cell surface membrane binder.

6.1.2 *Pseudomonas aeruginosa cytotoxin*

An exotoxin secreted by certain strains of *P. aeruginosa,* originally named leucocidin (Scharmann 1976a, b; Scharmann et al. 1975) and subsequently re-designated *P. aeruginosa* cytotoxin (Lutz et al. 1982), elicits detrimental effects on nucleated mammalian cells, including granulocytes (Baltch et al. 1985) and cells of the distal tubule of the kidney (Weiner and Reinacher 1982). The toxin is lethal in mice. The relevance of these findings to bacterial pathogenicity is under investigation. The toxin has been reported to be of M_r 23 000–25 000, and to have a pI of 5.0–6.4. At high doses (of the order of 0.1–1.0 mg/ml) the toxin appears to lyse rabbit erythrocytes, but not erythrocytes from sheep, horse, or cow. The existence of specific membrane binder molecules, of which these findings are suggestive, has not yet been established.

Evidence for pore formation by the toxin derives from studies of permeability changes in Ehrlich mouse ascites tumor cells and in cultured pulmonary endothelial cells (Lutz et al. 1982; Suttorp et al. 1985b). Both studies indicate that the toxin generates small, non-selective transmembrane pores of 0.5–1.5 nm diameter. Studies using cultured endothelial cells have also demonstrated a Ca^{2+}-dependent stimulation of the arachidonic acid cascade similar to that found with α-toxin. Physicochemical and immunological data on the membrane-bound form of the toxin are not yet available. It is also not clear whether the toxin pores exhibit a recognizable ultrastructure, and their inclusion in this section is currently based on suspicion only.

6.1.3 Aeromonas Cytolysin

The gram-negative pathogen *Aeromonas hydrophila* produces a cytotoxic agent that appears to form transmembrane pores of about 1.5—2.0 nm diameter in target erythrocytes (Howard and Buckley 1982). The native protein may initially bind to a cell-surface glycoprotein. Detailed data on the membrane-bound form of the toxin are not yet available.

6.2   Other Pore Formers

6.2.1   Gramicidin A

Gramicidin A is a polypeptide of 15 residues, all with hydrophobic side chains, produced by *Bacillus brevis*. It inserts spontaneously and induces ion-permeability increases in artificial lipid and cell membranes (Tosteson et al. 1968; Haydon and Hladky 1972). It appears that monomers of the toxin bind to target lipid bilayers and then form transient channels by an unstable dimerization process (Urry 1971, 1972; Hladky and Haydon 1972). The dimer channel is estimated to have an effective diameter of only about 0.4 nm and exhibits a charge and ion-size dependent selectivity, so permeability for small cations is governed by the ionic mobilities of the individual ions in water (Myers and Haydon 1972).

6.2.2   Pore-Forming Proteins From Sea Anemone, Fungi, and Amoeba

Several species of sea anemone produce related cytolytic toxins, of which the cytolysin of *Stoichactis helianthus* has been studied in some detail. Studies using planar lipid membranes indicate that this toxin aggregates to form oligomeric transmembrane pores (Michaels 1979). Bioimmunochemical and ultrastructural data on the pore are not available.

Entamoeba histolytica also produces a pore-forming cytolysin, the mode of action of which has been studied in planar lipid membranes (Young et al. 1982; Lynch et al. 1982). This protein appears to form pores in monomeric form. Once inserted into a planar lipid membrane, pore conductivity is negatively affected by addition of pronase to both *cis* and *trans* sides of the membrane (Young et al. 1982); hence, the protein appears to span the membrane and the pore may be destroyed by the action of proteases. A possible role for the cytolysin in the pathogenesis of amoebic dysentery has been discussed (Lynch et al. 1982). The M_r in SDS of the toxin studied by Lynch et al. appears to be only about 13 000. We assume that the toxins studied by both groups are identical, although formal proof to this effect is not available at present.

Phallolysin is a cytolysin of M_r 34 000 from *amanita phalloides*, with an amino acid composition that is remarkably similar to that of staphylococcal α-toxin (Faulstich et al. 1983). Further similarities in the physicochemical properties of α-toxin and phallolysin have been emphasized, including the existence of multpile forms, p^I values, and thermolability. However, the latter criteria should be critically reexamined in the light of the known oligomerizing properties of α-toxin. Although phallolysin appears to form pores in the membrane, little is known of the mode of pore formation and of the structure of the pore-forming molecules. Ultrastructural lesions such as those formed by α-toxin have not been reported.

7 Concluding Remarks

The concept of damage to mammalian cell membranes by pore-forming proteins is today well documented. New poreformers are being increasingly recognized, and the spectrum of producers of pore formers is already known to range from prokaryonts to plants and lower parasitic and mammalian cells. It has been the object of this review to summarize current knowledge in this steadily growing field, and to place the concept of pore formation into a more general perspective with regard to its biological significance. Although the best-studied proteinaceous pore formers have been represented by prototypes of the oligomerizing group with typical ultractructure, it should be stressed that many, and perhaps even the majority of, other proteinaceous pores may not exhibit characteristic ultrastructures that are detectable by present-day electron-microscopic techniques. Irrespective of whether a pore former belongs to the oligomerizing group, it must be endowed with the unique property of undergoing a hydrophilic-amphiphilic transition upon gaining contact with the target lipid bilayer, and be able subsequently to insert into the membrane. Despite the rapid progress that is being made at a molecular genetic level, with the amino acid sequences of several pore formers already available, a molecular model for the insertion process has not yet been determined for any pore former.

Seen in broader perspective, the principle of hydrophilic-amphiphilic transition of proteins appears essential for a variety of other important biological phenomena. Examples include the pH-dependent transition of certain exotoxin fragments (e.g., fragment B of diphtheria toxin; Kagan et al. 1981) into amphiphilic moieties; this is believed to be instrumental in the process of transport of intracellularly active fragments (e.g., fragment A of diphtheria toxin) across endosomal membranes (cf. Sandvig and Olsnes 1981). Viral fusion proteins undergo pH-dependent conformational transitions, exposing apolar areas that are probably associated with fusogenic activity and with extrusion of the viruses from endosomes (cf. Helenius et al. 1980; Skehel et al. 1982). Some primarily water-soluble microsomal proteins must pass into and sometimes through mitochondrial membranes, and these processes may similarly be connected with an exposure of lipid-binding sites (cf. Sabatini et al. 1982). Whether these and related phenomena derive from the existence of polypeptide domains comparable to "signal sequences" (Blobel 1980) remains to be explored. Finally, structural similarities may emerge between membrane-damaging pore formers (e.g., staphylococcal α-toxin), and naturally occurring pore formers such as the porins that are present in bacterial and mitochondrial outer membranes (for example see Benz 1986), and gap junctions, the acetylcholine receptor and sodium channels of mammalian cells (for recent overview, see Unwin 1986).

Acknowledgments. Original research from the authors' laboratories referred to in this article was supported by the Deutsche Forschungsgemeinschaft and the Verband der Chemischen Industrie e.V. (Fonds der Chemischen Industrie).

References

Adler S, Baker PJ, Pritzl P, Couser WG (1984) Detection of teminal complement components in experimental immune glomerular injury. Kidney Int 26:830–837

Adler S, Baker PJ, Johnson RJ, Ochi RF, Pritzl P, Couser WG (1986) Complement membrane attack omplex stimulates production of reaktive oxygen metabolites by cultured rat mesangial cells. J Clin Invest 77:762–767

Ahnert-Hilger G, Bhakdi S, Gratzl M (1985) Minimal requirements for exocytosis: a study using PC 12 cells permeabilized with staphylococcal α-toxin. J Biol Chem 260:12730–12734

Alouf JE (1980) Streptocococcal toxins (streptolysin O, streptolysin S, erythrogenic toxin). Pharmacol Ther 11:661–717

Alouf JE, Geoffroy C (1979) Comparative effects of cholesterol and thiocholesterol on streptolysin O. FEMS Microbiol Lett 6: 413–416

Alouf JE, Raynaud M (1973) Purification and some properties of streptolysin O. Biochimie 55:1187–1193

Alving CR, Habig WH, Urban KA, Hardegree MC (1979) Cholesterol-dependent tetanolysin damage to liposomes. Biochim Biophys Acta 551:224–228

Amiguet TP, Brunner J, Tschopp J (1985) The membrane attack complex of complement: lipid insertion of tubular and non-tubular polymerized C9. Biochemistry 24:7328–7334

Andersen BR, Amirault JJ (1976) Decreased E-rosette formation following streptolysin O treatment. Proc Soc Exp Biol Med 153:405–407

Andersen BR, Cone R (1974) Inhibition of human lymphocyte blast transformation by streptolysin O. J Lab Clin Med 84:241–248

Arbuthnott JP, Freer JH, Bernheimer AW (1967) Physical states of staphylococcal α-toxin. J Bacteriol 94:1170–1177

Arbuthnott JP, Freer JH, Billcliffe B (1973) Lipid-induced polymerization of staphylococcal α-toxin. J Gen Microbiol 75:309–319

Bader MF, Thierse D, Aunis D, Ahnert-Hilger G, Gratzl M (1986) Characterization of hormone and protein release from α-toxin permeabilized chromaffin cells in primary culture. J Biol Chem 261:5777–5783

Baltch AL, Hammer MC, Smith RP, Obrig TG, Conroy JV, Bishop MB, Egy MA, Lutz F (1985) Effects of *Pseudomonas aeruginosa* cytotoxin on human serum and granulocytes and their microbicidal, phagocytic and chemotactic functions. Infect Immun 48:498–506

Barnes DW, Silnutzer J (1983) Isolation of human serum spreading factor. J Biol Chem 258:12548–12552

Barnes DW, Silnutzer J, See C, Shaffer M (1983) Characterization of human serum spreading factor with monoclonal antibody. Proc Natl Acad Sci USA 80:1362–1366

Bashford CL, Alder GM, Patel K, Pasternak CA (1984) Common action of certain viruses, toxins and activated complement: pore formation and its prevention by extracellular Ca^{2+}. Biosci Rep 4:797–805

Behnke O, Tranum-Jensen J, van Deurs B (1986a) Filipin as a cholesterol probe. I Morphology of filipin-cholesterol interaction in lipid model systems. Eur J Cell Biol 35:189–199

Behnke O, Tranum-Jensen J, van Deurs B (1986b) Filipin as a cholesterol probe. II Filipin-cholesterol interaction in red blood cell membranes. Eur J Cell Biol 35:200–215

Benz R (1986) Porin from bacterial and mitochondrial outer membranes. CRC Crit Rev Biochem 19:145–190

Bernheimer AW (1974) Interactions between membranes and cytolytic bacterial toxins. Biochim Biophys Acta 344:27–50

Bernheimer AW, Rudy B (1986) Interactions between membranes and cytolytic peptides. Biochim Biophys Acta 864:123–141

Bhakdi S and Muhly M (1983) A simple immunoradiometric assay for the terminal SC5b-9 complex of human complement. J Immunol Meth 57:283–289

Bhakdi S, Roth W (1981) Fluid-phase SC5b-8 complex of human complement: generation and isolation from serum. J Immunol 127:576–582

Bhakdi S, Tranum-Jensen J (1978) Molecular nature of the complement lesion. Proc Natl Acad Sci USA 75:5655–5659

Bhakdi S, Tranum-Jensen J (1979) Evidence for a two-domain structure of the terminal membrane C5b-9 complex of human complement. Proc Natl Acad Sci USA 76:5872–5876

Bhakdi S, Tranum-Jensen J (1980) Re-incorporation of the terminal C5b-9 complement complex into lipid bilayers: formation and stability of reconstituted liposomes. Immunology 41:737–742

Bhakdi S, Tranum-Jensen J (1981) Molecular weight of the membrane C5b-9 complex of human complement: characterization of the terminal complex as a C5b-9 monomer. Proc Natl Acad Sci USA 78:1818–1822

Bhakdi S, Tranum-Jensen J (1983) Membrane damage by complement. Biochim Biophys Acta 737:343–372

Bhakdi S, Tranum-Jensen J (1984a) On the cause and nature of C9-related heterogeneity of C5b-9(m) complexes generated on erythrocyte targets through the action of whole human serum. J Immunol 133:1453–1463

Bhakdi S, Tranum-Jensen J (1984b) Mechanism of complement cytolysis and the concept of channel-forming proteins. Phil Trans R Soc Lond B 306:311–324

Bhakdi S, Tranum-Jensen J (1985) Complement activation and attack on autologous cells induced by streptolysin-O. Infect Immun 48:713–719

Bhakdi S, Tranum-Jensen J (1986) C5b-9 assembly: average binding of one C9 molecule to C5b-9 without poly-C9 formation generates a stable transmembrane pore. J Immunol 136:2999–3005

Bhakdi S, Bjerrum OJ, Rother U, Knüfermann H, Wallach DFH (1975a) Immuno-chemical analyses of membrane-bound complement: detection of the terminal complement complex and its similarity to intrinsic erythrocyte membrane proteins. Biochim Biophys Acta 406:21–35

Bhakdi S, Knüfermann H, Wallach DFH (1975b) Two-dimensional separation of erythrocyte membrane proteins. Biochim Biophys Acta 394: 550–557

Bhakdi S, Ey P, Bhakdi-Lehnen B (1976) Isolation of the terminal complement complex from target sheep erythrocyte membranes. Biochim Biophys Acta 419:445–457

Bhakdi S, Bhakdi-Lehnen B, Bjerrum OJ (1977) Detection of amphiphilic proteins and peptides in complex mixtures: charge-shift crossed immunoelectrophoresis and two-dimensional charge-shift electrophoresis. Biochim Biophys Acta 470: 35–44

Bhakdi S, Bjerrum OJ, Bhakdi-Lehnen B, Tranum-Jensen J (1978) Complement lysis: evidence for an amphiphilic nature of the membrane C5b-9 complex. J Immunol 121:2526–2532

Bhakdi S, Tranum-Jensen J, Klump O (1980) The terminal membrane C5b-9 complex of human complement: evidence for the existence of multible protease-resistant polypeptides that form the trans-membrane complement channel. J Immunol 124:2451–2457

Bhakdi S, Füssle R, Tranum-Jensen J (1981) Staphylococcal α-toxin: oligomerisation of hydrophilic monomers to form amphiphilic hexamers induced through contact with deoxycholate detergent micelles. Proc Natl Acad Sci USA 78: 5475–5479

Bhakdi S, Füssle R, Utermann G, Tranum-Jensen J (1983a) Binding and partial inactivation of S. aureus α-toxin by human plasma low density lipoprotein. J Biol Chem 258:5899–5904

Bhakdi S, Muhly M, Roth M (1983b) Isolation of specific antibodies to complement components. Methods Enzymol 93:409–420

Bhakdi S, Muhly M, Füssle R (1984a) Membrane damage by staphylococcal α-toxin: correlation between toxin-binding and hemolytic activity. Infect Immun 46: 318–323

Bhakdi S, Roth M, Sziegoleit A, Tranum-Jensen J (1984b) Streptolysin-O: isolation and identification of two hemolytic forms. Infect Immun 46:394–400

Bhakdi S, Tranum-Jensen J, Sziegoleit A (1985) Mechanism of membrane damage by streptolysin-O. Infect Immun 47:52–60

Bhakdi S, Mackman N, Nicaud JM, Holland IB (1986) E. coli hemolysin damages target cell membranes by generating trans-membrane. pores. Infect Immun 52:63–69

Biesecker G (1983) Biology of disease. Membrane attack complex of complement as a pathologic mediator. Lab Invest 49:237–249

Biesecker G, Müller-Eberhard HJ (1980) The ninth component of human complement: purification and physicochemical characterization. J Immunol 124: 1291–1296

Biesecker G, Katz S, Koffler D (1981) Renal localization of the membrane attack complex in systemic lupus erythematosus nephritis. J Exp Med 154:1779–1794

Biesecker G, Gerard G, Hugli TE (1982a) An amphiphilic structure of the ninth component of human complement. J Biol Chem 257:2584–2590

Biesecker G, Lavin L, Ziskind M, Koffler D (1982b) Cutaneous localization of the membrane attack complex in discoid and systemic lupus erythematosus. N Engl J Med 306:264–270

Biesecker G, Noble B, Andres GA, Koffler D (1984) Immunopathogenesis of Heymann's nephritis. Clin Immunol Immunopathol 33:333–338

Blobel G (1980) Intracellular protein topogenesis. Proc Natl Acad Sci USA 77: 1496–1500

Blumenthal R, Millard PJ, Henkart MP, Reynolds CW, Henkart PA (1984) Liposomes as targets for granule cytolysin from cytotoxic LGL tumors. Proc Natl Acad Sci USA 81:5551–5555

Borsos T, Dourmashkin RR, Humphrey JH (1964) Lesions in erythrocyte membranes caused by immune hemolysis. Nature 202:251–254

Boyer JT, Gall EP, Norman ME, Nilsson UR, Zimmermann TS (1975) Hereditary deficiency of the seventh component of complement. J Clin Invest 56:905–913

Boyle MDP, Borsos T (1979) Studies on the terminal stages of immune hemolysis. V. Evidence that not all complement-produced channels are equal. J Immunol 123:71–76

Boyle MDP, Langone JJ, Borsos T (1978) Studies on the terminal stages of immune hemolysis. III Distinction between the insertion of C9 and the formation of a transmembrane channel. J Immunol 120: 1721–1725

Boyle MDP, Gee AP, Borsos T (1979) Studies on the terminal stages of immune hemolysis. VI. Osmotic blocks of differing Stokes' radii detect complement-induced transmembrane channels of differing size. J Immunol 123:77–82

Boyle MDP, Gee AP, Borsos T (1981) Heterogeneity in the size and stability of transmembrane channels produced by whole complement. Clin Immunol Immunopathol 20:287–295

Bremm KD, König W, Pfeiffer P, Rauschen I, Theobald K, Thelestam M, Alouf JE (1985) Effect of thiol-activated toxins (streptolysin O, alveolysin, and theta toxin) on the generation of leukotrienes and leukotriene-inducing and -metabolizing enzymes from human polymorphonuclear granulocytes. Infect Immun 50: 844–851

Buckelew AR, Colaccico G (1971) Lipid monolayers. Interaction with staphylococcal α-toxin. Biochim Biophys Acta 233:7–16

Buckingham L, Duncan JL (1983) Approximate dimensions of membrane lesions produced by streptolysin-S and streptolysin-O. Biochim Biophys Acta 729: 115–122

Campbell AK, Morgan BP (1985) Monoclonal antibodies demonstrate protection of polymorphonuclear leukocytes against complement attack. Nature 317: 164–166

Carney DJ, Koski CL, Shin ML (1985) Elimination of terminal complement intermediates from the plasma membrane of nucleated cells: the rate of disappearance differs for cells carrying C5b-7 or C5b-8 or a mixture of C5b-8 with a limited number of C5b-9. J Immunol 134:1804–1809

Carney DF, Hammer CH, Shin ML (1986) Elimination of terminal complement complexes in the plasma membrane of nucleated cells: influence of extracellular Ca^{2+} and association with cellular Ca^{2+}. J Immunol 137:263–270

Cassidy P, Six HR, Harshman S (1976) Studies on the binding of staphylococcal ^{125}I-labelled α-toxin to rabbit erythrocytes. Biochemistry 15:2348–2355

Cavalieri SJ, Bohach GA, Snyder IS (1984) *Escherichia coli* α-hemolysin: characteristics and probable role in pathogenicity. Microbiol Rev 48:326–343

Cheng KH, Wiedmer T, Sims PJ (1985) Fluorescence resonance energy transfer study of the associative state of membrane-bound complexes of complement proteins C5b-8. J Immunol 135:459–464

Cosyns JP, Kazatchkine MD, Bhakdi S, Mandet C, Grossetete J, Hinglais N, Bariety J (1986) Immunohitochemical analysis of C3-cleavage fragments, factor H, and the C5b-9 terminal complex of complement in de novo membranous glomeru-lonephritis occurring in patients with renal transplant. Clin Nephrol 26:203—209

Couser WG, Baker PJ, Adler S (1985) Complement and the direct mediation of immune glomerular injury: a new perspective. Kidney Int 28:879—890

Cowell JL, Bernheimer AW (1977) Antigenetic relationships among thiol-activated cytolysins. Infect Immun 16:397—399

Cowell JL. Kim KS, Bernheimer AW (1978) Alteration by cereolysin of the structure of cholesterol-containing membranes. Biochim Biophys Acta 507:230—241

Cramer WA, Dankert JR, Uratani Y (1983) The membrane channel-forming bacteri-ocidal protein, colicin E1. Biochim Biophys Acta 737:173—193

Cybulsky AV, Rennke HG, Feintzeig ID, Salant DJ (1986) Complement-induced glomerular epithelial cell injury. J Clin Invest 77:1096—1107

Dahl MV, Falk RJ, Carpenter R, Michael AF (1984) Deposition of the membrane attack complex of complement in bullous pemphigoid. J Invest Dermatol 82:132—135

Dahlbäck B, Podack ER (1985) Characterization of human S protein, an inhibitor of the membrane attack complex of complement. Demonstration of a free reactive thiol group. Biochemistry 24:2368—2374

Dalmasso AP, Bension BA (1981) Lesions of different functional size produced by human and guinea pig complement in sheep red cell membranes. J Immunol 127:2214—2218

Dankert JR, Esser AF (1985) Protelytic modification of human complement protein C9: loss of poly (C9) and circular lesion formation without impairment of function. Proc Natl Acad Sci USA 82:2128—2132

DeHeer E, Daha MR, Bhakdi S, Bazin H, van Es LA (1985) Possible involvement of terminal complement complex in active Heymann nephritis. Kidney Int 22: 388—393

Dennert G, Podack ER (1983) Cytolysis by H-2 specific T killer cells: assembly of tubular complexes on target membranes. J Exp Med 157:1483—1495

Densen P, Brown EJ, O'Neill GJ, Tedesco F, Clark RA, Frank MM, Webb D, Myers J (1983) Inherited deficiency of C8 in a patient with recurrent meningococcal infections: further evidence for a dysfunctional C8 molecule and nonlinkage to the HLA system. J Clin Immunol 3:90—99

DiScipio RG, Gehring MR, Podack ER, Chen Chen Kan, Hugli TE, Fey GH (1984) Nucleotide sequence of cDNA and derived amino acid sequence of human complement component C9. Proc Natl Acad Sci USA 81:7298—7302

Dourmashkin RR, Rosse WF (1966) Morphologic changes in the membranes of red blood cells undergoing hemolysis. Am J Med 41:699—710

Dourmashkin RR, Deteix P, Simone CB, Henkart PA (1980) Electron microscopic demonstration of lesions on target cell membranes associated with antibody-dependent cytotoxicity. Clin Exp Immunol 42:554—560

Duncan JL (1974) Characteristics of streptolysin O hemolysis: kinetics of hemo-globin and [86]rubidium release. Infect Immun 9:1022—1027

Duncan JL, Buckingham L (1977) Effects of streptolysin O on transport of amino acids, nucleosides and glucose analogs in mammalian cells. Infect Immun 18: 688—693

Duncan JL, Schlegel R (1975) Effect of streptolysin O an erythrocyte membranes, liposomes, and lipid dispersions. A protein-cholesterol interaction. J Cell Biol 67:160—173

Ehlenberger AG, Nussenzweig V (1977) The role of membrane receptors for C3b and C3d in phagocytosis. J Exp Med 145:357—371

Engel AG, Bisecker G (1982) Complement activation in muscle fiber necrosis: demonstration of the membrane attack complex of complement in necrotic fibers. Ann Neurol 12:289—296

Esser AF (1983) Interactions between complement proteins and biological and model membranes. In: Chapman CC (ed) Biological membranes, vol 4. Academic, New York, pp 277–322

Falk RJ, Dalmasso AP, Kim Y, Tasi Ch, Scheinman JI, Gewurz H, Michael AF (1983) Neoantigen of the polymerized ninth component of complement. Characterization of a monoclonal antibody and immunohistochemical localization in renal disease. J Clin Invest 72:560–573

Faulstich H, Bühring HJ, Seitz J (1983) Physical properties and function of phallolysin. Biochemistry 22:4574–4580

Fearon DT (1979) Regulation of the amplifcation C3 convertase of human complement by an inhibitory protein isolated from human erythrocyte membranes. Proc Natl Acad Sci USA 76:5867–5871

Fearon DT (1980) Identification of the membrane glycoprotein that is the C3b receptor of the human erythrocyte, polymorphonuclear leukocyte, B-lymphocyte and monocyte. J Exp Med 152:20–30

Fearon DT (1985) Human complement receptors for C3b (CR1) and C3d (CR2). J Invest Dermatol 85:533–569

Fearon DT, Kaneko I, Thompson G (1981) Membrane distribution and adsorptive endocytosis by C3b receptors on human polymorphonuclear leukocytes. J Exp Med 153:1615–1628

Felmlee T, Pellet S, Welch R (1985) Nucleotide sequence of an *Escherichia coli* chromosomal hemolysin. J Bacteriol 163:94–105

Fischer E, Kazatchkine MD, Mecarelli-Halbwuchs L (1984) Protection of the classical and alternative complement pathway C3-convertases, stabilized by nephritic factors, from decay by the human C3b receptor. Eur J Immunol 14:1111–1114

Freer JH, Arbuthnott JP, Bernheimer AW (1968) Interaction of staphylococcal α-toxin with artificial and natural membranes. J Bacteriol 95:1153–1168

Freer JH, Arbuthnott JP, Billcliffe B (1973) Effects of staphylococcal α-toxin on the structure of erythrocyte membranes: a biochemical and freeze-etching study. J Gen Microbiol 75:321–332

Fried FA, Vermeulen CW, Ginsburg MJ, Cone CM (1971) Etiology of pyelonephritis: further evidence associating the production of experimental pyelonephritis with hemolysis in *Escherichia coli*. J Urol 106:351–354

Füssle R, Bhakdi S, Sziegoleit A, Tranum-Jensen J, Kranz T, Wellensiek HJ (1981) On the mechanism of membrane damage by *S. aureus* α-toxin. J Cell Biol 91: 83–94

Giavedonie EB, Chow YM, Dalmasso AP (1979) The functional size of the primary complement lesion in resealed erythrocyte membrane ghosts. J Immunol 122: 240–245

Gigli I, Nelson RA (1968) Complement dependent immune phagocytosis. Exp Cell Res 51:45–67

Goebel W, Hedgepeth J (1982) Cloning and functional characterization of the plasmid-encoded hemolysin determinant of *Escherichia coli*. J Bacteriol 151: 1290–1298

Goldlust MB, Shin HS, Hammer CH, Mayer MM (1974) Studies of complement complex C5b6 eluted from EAC-6; reaction of C5b,6 with EAC 4b, 3b and evidence on the role of C2a and C3b in the activation of C5. J Immunol 113: 998–1007

Gonzalez-Carrero MI, Zabala JC, de la Cruz F, Oritz JM (1985) Purification of α-haemolysin from an overproducing *E. coli* strain. Mol Gen Genet 199:106–110

Gray GS, Kehoe M (1984) Primary sequence of the α-toxin gene from staphylococcus aureus wood 46. Infect Immun 46:615–618

Green H, Barrow P, Goldberg B (1959) Effect of antibody and complement on permeability control in ascites tumor cells and erythrocytes. J Exp Med 110: 699–712

Groggel GC, Adler S, Rennke HG, Couser WG, Salant DJ (1983) Role of the terminal complement pathway in experimental membranous nephropathy in the rabbit. J Clin Invest 72:1948–1957

Groggel GC, Salant DJ, Darby C, Rennke HG, Couser WG (1985) Role of terminal complement pathway in the heterologous phase of antiglomerular basement membrane nephritis. Kidney Int 27:643–651

Gupta RK, Srimal RC (1979) Effect of intraventricular administration of streptolysin O on the electroencephalogram of rabbits. Toxicon 17:321–325

Haberman E (1972) Bee and wasp venoms. Science 177:314–322

Hacker J, Hughes C, Hof H, Goebel W (1983) Cloned hemolysin genes from *Escherichia coli* that cause urinary tract infection determine different levels of toxicity in mice. Infect Immun 42:57–63

Haeney MR, Thompson RA, Faulkner J, Mackintosh P, Ball AP (1980) Recurrent bacterial meningitis in patients with genetic defects of terminal complement components. Clin Exp Immunol 40:16–24

Halbert SP, Bircher R, Dahle E (1963a) Studies on the mechanism of the lethal toxic action of streptolysin O and the protection by certain serotonin drugs. J Lab Clin Med 437–452

Halbert SP, Dahle E, Keatinge S, Bircher R (1963b) Studies on the role of potassium ions in the lethal toxicity of streptolysin O. In: Raskova H (ed) Recent advances in pharmacology toxins. Pergamon, Oxford, pp 439–453

Halpern BN, Rahman S (1968) Studies on the cardiotoxicity of streptolysin O. Br J Pharmacol 32:441–452

Hammer CH, Nicholson A, Mayer MM (1975) On the mechanism of cytolysis by complement: evidence on insertion of C5b and C7 subunits of the C5b-6, 7 complex into the phospholipid bilayer of erythrocyte membranes. Proc Natl Acad Sci USA 72:5076–5080

Hammer CH, Wirtz GH, Renfer L, Gresham HD, Tack BF (1981) Large scale isolation of functionally active components of the human complement system. J Biol Chem 256:3995–4006

Hänsch GM, Hammer CH, Vanguri P, Shin ML (1981) Homologous species restriction in lysis of erythrocytes by terminal complement proteins. Proc Natl Acad Sci USA 78:5118–5122

Hänsch GM, Seitz M, Martinotti G, Betz MM, Rauterberg EW, Gemsa D (1984) Macrophages release arachidonic acid, prostaglandin E_2, and thromboxane in response to late complement components. J Immunol 133:2145–2150

Hänsch GM, Gemsa D, Resch K (1985) Induction of prostanoid synthesis in human platelets by the late complement components C5b-9 and channel-forming antibiotic nystatin: inhibition of the eacylation of liberated arachidonic acid. J Immunol 135:1320–1324

Harshman S (1979) Action of staphylococcal α-toxin on membranes: some recent advances. Mol Cell Biochem 23:142–152

Harshman S, Sugg N (1985) Effect of calcium ions on staphylococcal alpha-toxin induced hemolysis of rabbit erythrocytes. Infect Immun 47:37–40

Harshman S, Burt AM, Robinson JP, Blankenship M, Harshman DL (1985) Disruption of myelin sheaths in mouse brain in vitro and in vivo by staphylococcal α-toxin. Toxican 23:801–806

Haydon DA and Hladky SB (1972) Ion transport across thin lipid membranes: a critical discussion of mechanisms in selected systems. Q Rev Biophys 5: 187–282

Hayman EG, Engvall E, A'Hearn E, Barnes D, Pierschbacher M, Ruoslahti E (1982) Cell attachment on replicas of SDS-polyacrylamide gels reveals two adhesive plasma proteins. J Cell Biol 95:20–23

Hayman EG, Pierschbacher MD, Öhgren Y, Ruoslahti E (1983) Serum spreading factor (vitronectin) is present at the cell surface and in tissues. Proc Natl Acad Sci USA 80:4003–4007

Helenius A, Simons K (1975) Solubilization of membranes by detergents. Biochim Biophys Acta 415:29−79

Helenius A, Simons K (1977) Charge-shift electrophoresis: a simple method for distinguishing between hydrophilic and amphiphilic proteins in detergent solution. Proc Natl Acad Sci USA 74:529−533

Helenius A, Kartenbeck J, Simons K, Fries E (1980) On the entry of Semliki Forest Virus into BHK-21 cells. J Cell Biol 84:404−420

Henkart P (1985) Mechanism of lymphocyte-mediated cytotoxicity. Annu Rev Immunol 3:31−58

Henkart P, Blumenthal R (1975) The interaction of lymphocytes with lipid bilayer membranes: a model for the lymphocyte-mediated lysis of target cells. Proc Natl Acad Sci USA 72:2789−2793

Henkart PA, Millard PJ, Reynolds CW, Henkart MP (1984) Cytolytic activity of purified cytoplasmic granules from cytotoxic rat LGL tumors. J Exp Med 160:75−93

Henney CS (1973) Studies on the mechanism of lymphocyte-mediated cytolysis. II The use of various target cell markers to study cytolytic events. J Immunol 110:73−84

Henney CS (1974) Estimation of the size of a T-cell-induced lytic lesion. Nature 249:456−458

Herbert D, Todd EW (1941) Purification and properties of a haemolysin produced by group a hemolytic streptococci (streptolysin O). Biochem J 35:1124−1139

Hewitt LF, Todd EW (1939) The effect of cholesterol and of sera contaminated with bacteria on the haemolysins produced by haemolytic streptococci. J Path Bacteriol 49:45−51

Hill JH, Ward PA (1971) The phlogistic role of C3 leukotactic fragments in myocardial infarcts of rats. J Exp Med 133:885−900

Hinglais N, Kazatchkine MD, Bhakdi S, Appay MD, Mandet C, Grossetete J, Bariety J (1986) Immunohistochemical study of the C5b-9 complex of complement in normal and diseased human kidneys: diversity in localization and potential in tissue damage. Kidney Int 30:399−410

Hladky SB, Haydon DA (1972) Ion ttransfer across lipid membranes in the presence of gramicidin A. I. Studies on the unit conductance channel. Biochim Biophys Acta 274:294−312

Holmes R (1967) Preparation from human serum of an alpha-one protein which induces the immediate growth of unadapted cells in vitro. J Cell Biol 32:297−308

Howard P, Buckley JT (1982) Membrane glycoprotein receptor and hole-forming properties of a cytolytic protein toxin. Biochemistry 21:1662−1667

Howard JG, Wallace KR, Wright GP (1953) The inhibitory effects of cholesterol and related sterols on haemolysis by streptolysin O. Br J Exp Pathol 34:174−180

Hu V, Nicholson-Weller A (1985) Enhanced complement-mediated lysis of type III paroxysmal nocturnal hemoglobinuria erythrocytes involves increased C9-binding and polymerization. Proc Natl Acad Sci USA 82:5520−5524

Hu V, Esser AF, Podack ER, Wisnieski BJ (1981) The membrane attack mechanism of complement: photolabeling reveals insertion of terminal proteins into target membrane. J Immunol 127:380−384

Hughes C, Hacker J, Robert A, Goebel W (1983) Hemolysin production as a virulence marker in symptomatic and asymptomatic urinary tract infections caused by *Escherichia coli*. Infect Immun 39:546−551

Hügli T, Müller-Eberhard HJ (1978) Anphylatoxins: C3a and C5a. Adv Immunol 26:1−55

Hugo F, Jenne D, Bhakdi S (1985) Monoclonal antibodies to neoantigens of the C5b-9 complex of human complement. Biosci Rep 5:649−658

Hugo F, Reichwein J, Arvand M, Krämer S, Bhakdi S (1986) Mode of transmembrane pore formation by streptolysin-O analysed with a monoclonal antibody. Infect Immun 54:641−645

Humphrey JH, Dourmashkin RR (1969) The lesions in cell membranes caused by complement. Adv Immunol 11:75–115

Imagawa DK, Osifchin NE, Paznekas WA, Shin ML, Mayer MM (1983) Consequences of cell membrane attack by complement: release of arachidonate and formation of inflammatory derivatives. Proc Nat Acad Sci USA 80:6647–6651

Imagawa DK, Osifchin NE, Ramm LE, Koga PG, Hammer CH, Shin HS, Mayer MM (1986) Release of arachidonic acid and formation of oxygenated derivatives after complement attack on macrophages: role of channel formation. J Immunol 136:4637–4643

Inoue K, Kinoshita T, Okada M, Akiyama Y (1977) Release of phospholipids from complement-mediated lesions on the furface structure of *Escherichia coli*. J Immunol 119:65–72

Ishida B, Wisnieski JB, Lavine H, Esser AF (1982) Photolabeling of a hydrophobic domain of the ninth component of human complement. J Biol Chem 257: 10551–10553

Jasin HE (1977) Absence of the eighth component of complement in association with systemic lupus erythematosus-like disease: J Clin Invest 60:709–715

Jeljaszewicz J, Szmigielski S, Hryniewicz W (1978) Biological effects of staphylococcal and streptococcal toxins. In: Jeljaszewicz J, Wadström T (eds) Bacterial toxins and cell membranes. Academic, New York, pp 185–227

Jenne D, Stanley KK (1985) Molecular cloning of S-protein, a link between complement coagulation and cell-substrate adhesion. EMBO J 4:3153–3157

Jenne D, Hugo F, Bhakdi S (1985) Interaction of complement S-protein with thrombin-antithrombin complexes: a role for the S-protein in haemostasis. Thromb Res 38:401–412

Johnson MK, Geoffroy C, Alouf JE (1980) The binding of cholesterol by sulfhydryl-activated cytolysins. Infect Immun 27:97–101

Jorgensen SE, Mulcahy PF, Wu GK, Louis CF (1983) Calcium accumulation in human and sheep erythrocytes that is induced by *Escherichia coli* hemolysin. Toxicon 21:717–727

Kagan BL, Finkelstein A, Colombini M (1981) Diphtheria fragment forms large pores in phospholipid bilayer membranes. Proc Natl Acad Sci USA 78:4950–4954

Kazatchkine J, Nydegger UE (1982) The human alternative complement pathway. Prog Allergy 30:193–222

Kehoe M, Timmis KN (1984) Cloning and expression in *E. coli* of the streptolysin-O determinant from streptococcus pyogenes. Infect Immun 43:804–810

Kinoshita T, Inoue K, Okada M, Akiyama Y (1977) Release of phospholipids from liposomal model membrane damaged by antibody and complement. J Immunol 119:65–72

Kinsky SC (1970) ANtibiotic interaction with model membranes. Annu Rev Pharmacol 10:119–142

Kinsky SC (1972) Antibody-complement interaction with lipid model membranes. Biochim Biophys Acta 265:1–23

Kissel JT, Mendell JR, Rammohan KW (1986) Microvascular deposition of complement membrane attack complex in dermatomyositis. N Engl J Med 314: 329–334

Koffler D, Biesecker G, Noble B, Andres GA, Martinze-Hernandez A (1983) Localization of the membrane attack complex in experimental immune complex glomerulonephritis. J Exp Med 157:1885–1905

Kolb WP, Müller-Eberhard HJ (1973) The membrane attack mechanism of complement. Verification of a stable C5-C9 complex in free solution. J Exp Med 138:438–451

Kolb WP, Müller-Eberhard HJ (1974) Mode of action of C9: adsorption of multiple C9 molecules to cell-bound C8. J Immunol 113:479–488

Kolb WP, Müller-Eberhard HJ (1975a) The membrane attack mechanism of complement. Isolation and subunit composition of the C5b-9 complex. J Exp Med 141:724–735

Kolb WP, Müller-Eberhard HJ (1975b) Neoantigens of the membrane attack complex of human complement. Proc Natl Acad Sci USA 72:1687–1691

Kolb WP, Haxby JA, Arroyave CM, Müller-Eberhard HJ (1972) Molecular analysis of the membrane attack mechanism of complement. J Exp Med 135:549–566

Konisky J (1982) Colicins and other bacteriocins with established modes of action. Annu Rev Microbiol 36:125–144

Kopp WC, Burrell R (1982) Evidence for antibody-dependent binding of the terminal complement component to alveolar basement membrane. Clin Immunol Immunopathol 23:10–21

Koski CL, Ramm LE, Hammer CH, Mayer MM, Shin ML (1983) Cytolysis of nucleated cells by complement cell death displays multi-hit characteristics. Proc Natl Acad Sci USA 80:3816–3820

Lachman RJ, Thompson RA (1970) Reactive lysis: the complement-mediated lysis of unsensitized cells. II. The characterization of activated reactor as C5b and the participation of C8 and C9. J Exp Med 131:643–657

Lachmann PJ, Munn EA, Weissmann G (1970) Complement-mediated lysis of liposomes produced by the reactive procedure. Immunology 19: 983–986

Lachmann PJ, Bowyer DE, Nichol P, Dawson RMC, Munn EA (1973) Studies on the terminal stages of complement lysis. Immunology 24:135–145

Latorre R, Alvarez O (1981) Voltage-dependent channels in planar bilayer membranes. Physiol Rev 61:77–150

Lauf PK (1975) Immunological and physiological characteristics of the rapid immune hemolysis of neuraminidase-treated sheep red cells produced by fresh guinea-pig serum. J Exp Med 142:974–988

Leddy JP, Frank MM, Gaither T, Baum J, Klemperer MR (1974) Hereditary deficiency of the sixth component of complement in man. I. Immunochemical, biologic and family studies. J Clin Invest 53:544–553

Lee TJ, Utsinger PD, Snyderman R, Yount WJ, Sparling PF (1978) Familial deficiency of the seventh component of complement associated with recurrent bacteremic infections due to neisseria. J Infect Dis 138:359–368

Lim D, Gewurz A, Lint TF, Ghaze M, Sepheri B, Gewurz H (1976) Absence of the sixth component of complement in a patient with repeated episodes of meningococcal meningitis. J Pediatrics 89:42–47

Linder R (1979) Heterologous immunoaffinity chromatography in the purification of streptolysin O. FEMS Microbiol Lett 5:339–342

Lint TF, Zeitz HJ, Gewurz H (1980) Inherited deficiency of the ninth component of complement in man. J Immunol 125:2252–2257

Logue GL, Rosse WF, Adams JP (1974) Mechanisms of immune lysis of red blood cells in vitro. I. Paroxysmal nocturnal hemoglobinuria cells. J Clin Invest 52: 1129–1137

Loirat C, Buriot D, Peltier AP, Birche P, Aujard Y, Griscelli C, Mathiew H (1980) Fulminant meningococcemia in child with hereditary deficiency of the seventh component of complement and proteinuria. Acta Paediatr Scand 69:553–557

Lutz F, Grieshaber S, Schmidt K (1982) Permeability changes of Ehrlich mouse ascites tumor cells induced by cytotoxin from Pseudomonas aeruginosa. Naunyn-Schmiedebergs Arch Pharmacol 320:78–80

Lynch EC, Rosenberg IM, Gitler C (1982) An ion-channel forming protein produced by Entamoeba histolytica. EMBO J 1:801–804

Mackman N, Holland IB (1984) Functional characterization of a cloned haemolysin determinant from E. coli of human origin, encoding information for the secretion of a 107K polypeptide. Mol Gen Genet 196:123–134

Mackman N, Nicaud JM, Gray L, Holland IB (1986) Secretion of hemolysin by E. coli. Curr Top Microbiol Immunol 125: 159–181

Maroko PR, Carpenter CB, Chiariello M, Fishbein MC, Radvany P, Knustman JD, Hale SL (1978) Reduction by cobra venom factor of myocardial necrosis after coronary artery occlusion. J Clin Invest 61:661–670

Masson D, Tschopp J (1985) Isolation of a lytic, pore-forming protein (perforin) from cytolytic T-lymphocytes. J Biol Chem 260:9069–9072

Martz E (1976) Early steps in specific tumor cell lysis by sensitized mouse T lymphocytes. II. Electrolyte permeability increase in the target cell membrane concomitant with programming for lysis. J Immunol 117:1023–1027

Mayer MM (1972) Mechanism of cytolysis by complement. Proc Natl Acad Sci USA 69:2954–2959

Mayer MM, Michaels DW, Ramm LE, Whitlow MB, Willoughby JB, Shin ML (1981) Membrane damage by complement. CRC Crit Rev Immunol 7:133–165

McCartney C, Arbuthnott JP (1978) Mode of action of membrane-damaging toxins produced by staphylococci. In: Jeljaszewicz J, Wadström T (eds) Bacterial toxins and cell membranes. Academic Press, New York, pp 89–127

McEwen BF, Arion WJ (1985) Permeabilization of rat hepatocytes with *staphylococcus aureus* α-toxin. J Cell Biol 100:1922–1929

McLeod B, Baker P, Behrends CL, Baker PJ, Gewurz M (1975a) Studies of the inhibition of C5b-initiated lysis (reactive lysis). Immunology 28:379–390

McLeod B, Baker P, Behrends F, Gewurz H (1975b) Studies on the inhibition of C5b-initiated lysis (reactive lysis). III. Characterization of the inhibitory activity C567-INH and its mode of action. Immunology 28:133–149

McManus LM, Kolb WP, Crawford MH, O'Rourke RA, Grover FL, Pinckard RN (1983) Complement localization in ischemic baboon myocardium. Lab Invest 48:447

Menestrina G (1986) Ionic channels formed by *Staphylococcus aureus* α-toxin: voltage-dependent inhibition by divalent and trivalent cations. J Mem Biol 90: 177–190

Michaels DW (1979) Membrane damage by a toxin from the sea anemone stoichactis helianthus. I. Formation of transmembrane channels in lipid bilayers. Biochim Biophys Acta 555:67–78

Michaels DW, Abramovitz AS, Hammer CH, Mayer MM (1976) Increased ion permeability of planar lipid bilayer membranes after treatment with the C5b-9 cytolytic attack mechanism of complement. Proc Natl Acad Sci USA 73:2852–2856

Millard PJ, Henkart MP, Reynolds CW, Henkart PA (1984) Purification and properties of cytoplasmic granules from cytotoxic rat LGL tumors. J Immunol 132: 3197–3202

Mitsui K, Sekiya T, Nozawa Y, Hase J (1979a) Alteration of human erythrocyte plasma membranes by perfringolysin O as revealed by freeze-fracture electron microscopy. Studies on *Colstridium per fringens* exotoxins V. Biochim Biophys Acta 554:68–75

Mitsui K, Sekiya T, Okamura S, Nozawa Y, Hase J (1979b) Ring formation of perfringolysin O as revealed by negative stain electron microscopy. Biochim Biophys Acta 558:307–313

Möllby R (1978) Bacterial phospholipases. In: Jeljaszewicz J, Wadström T (eds) Bacterial toxins and cell membranes. Academic, London, pp 367–424

Möllby R (1983) Isolation and properties of membrane-damaging toxins. In: Easmon CSF, Adlam C (eds) Staphlococci and staphylococcal infections, vol 2. Academic, London, pp 619–669

Monahan JB, Sodetz JM (1980) Binding of the eighth component of human complement to the soluble cytolytic complex is mediated by its β subunit. J Biol Chem 255:10579–10582

Monahan JB, Sodetz JM (1981) Role of the β-subunit in the interaction of the eigth component of human complement with the membrane bound cytolytic complex. J Biol Chem 256:3433—3441

Montal M (1974) Formation of bimolecular membranes from lipid monolayers. Methods Enzymol 32:545—554

Morgan BP, Campbell AK (1985) The recovery of human polymorphonuclear leucocytes from sublytic complement attack is mediated by changes in intracellular free calcium. Biochem J 231:205—208

Morgan BP, Luzio JP, Campbell AK (1984) Inhibition of complement-induced ^{14}C sucrose release by intracellular and extracellular monoclonal antibodies to C9: evidence that C9 is a transmembrane protein. Biochem Biophys Res Comm 118:616—627

Müller-Eberhard HJ (1975) Complement. Annu Rev Biochem 44:697—723

Müller-Eberhard HJ (1984) The membrane attack complex. Springer Semin Immunopathol 7:93—141

Myers VB, Haydon DA (1972) Ion transfer across lipid membranes in the presence of gramicidin A. II. The ion selectivity. Biochim Biophys Acta 274:313—322

Nemerow GR, Gewurz H, Osofsky SG, Lint TF (1978) Inherited deficiency of the seventh component of complement associated with nephritis. J Clin Invest 61:1602—1610

Nicaud JM, Mackman N, Gray L, Holland IB (1985) Regulation of haemolysin synthesis in E. coli determined by Hly genes of human origin. Mol Gen Genet 199:111—116

Nicholson-Weller A, March JP, Rosenfeld SI, Austen KF (1983) Affected erythrocytes of patients with paroxysmal nocturnal hemoglobinuria are deficient in the complement regulatory protein decay accelerating factor. Proc Natl Acad Sci USA 80:5066—5070

Niedermeyer W (1985) Interaction of streptolysin-O with biomembranes: Kinetic and morphological studies on erythrocyte membranes. Toxicon 23:425—439

Norman AW, Spielvogel AM, Wong RC (1976) Polyene antibiotic-sterol interaction. Adv Lipid Res 14:127—170

Ofek I, Bergner-Rabinowitz S, Ginsburg I (1972) Oxygen-stable hemolysins of group A streptococci. VIII. Leukotoxic and antiphagocytic effects of streptolysins S and O. Infect Immun 6:459—464

Ohanian SH, Schlager SI (1981) Humoral immune killing of nucleated cells: mechanisms of complement-mediated attack and target cell defense. CRC Crit Rev Immunol 2:165—209

Packman CH, Rosenfeld SI, Jenkins DE, Thiem PA, Leddy JP (1979) Complement lysis of human erythrocytes. Differing susceptibility of two types of paroxysmal nocturnal hemoglobinuria cells to C5b-9. J Clin Invest 69:428—433

Pangburn MK, Schreiber RD, Müller-Eberhard HJ (1983) Deficiency of an erythrocyte membrane protein with complement regulatory activity in paroxysmal nocturnal hemoglobinuria. Proc Natl Acad Sci USA 80:5430—5434

Parker CJ, Wiedmer TW, Sims PJ, Rosse WF (1985) Characterization of the complement sensitivity of paroxysmal nocturnal hemoglobinuria erythrocytes. J Clin Invest 75:2074—2084

Parra G, Platt JL, Falk RJ, Rodriguez-Iturbe B, Michael AF (1984) Cell populations and membrane attack complex in glomeruli of patients with post-streptococcal glomerulonephritis: identification using monoclonal antibodies by indirect immunoflurescence. Clin Immunol Immunopathol 33:324—332

Parrisius J, Bhakdi S, Roth M, Tranum-Jensen J, Goebel W, Seeliger HRP (1986) Production and non-production of listeriolysin by beta-hemolytic strains of Listeria monocytogenes. Infect Immun 51:314—319

Perkinson DT, Baker PJ, Couser WG, Johnson RF, Adler S (1985) Membrane attack complex deposition in experimental glomerular injury. Am J Pathol 120: 121–128

Petersen BH, Fraham JA, Brooks GF (1976) Human deficiency of the eight component of complement. J Clin Invest 57:283–288

Pinckard RN, O'Rourke RA, Crawford MH, Grover FS, McManus LM, Ghidoni JJ, Storrs SB, Olson MS (1980) Complement localization and mediation of ischemic injury in baboon myocardium. J Clin Invest 66:1050–1056

Podack ER (1984) Molecular composition of the tubular structure of the membrane attack complex of complement. J Biol Chem 259:8641–8647

Podack ER, Dennert G (1983) Cell mediated cytolysis: assembly of two types of tubules with putative cytolytic function by cloned natural killer cells. Nature 302:442–445

Podack ER, Konigsberg PJ (1984) Cytolytic T cell granules. Isolation, biochemical and functional characterization. J Exp Med 160:695–710

Podack ER, Müller-Eberhard HJ (1978) Binding of deoxycholate, phosphatidyl choline vesicles, lipoprotein and of the S-protein to complexes of terminal complement components. J Immunol 121:1025–1030

Podack ER, Tschopp J (1982a) Circular polymerization of the ninth component of complement (poly C9): ring closure of the tubular complex confers resistance to detergent dissociation and to proteolytic degradation. J Biol Chem 257: 15204–15212

Podack ER, Tschopp J (1982b) Polymerization of the ninth component of complement (C9): formation of poly C9 with a tubular ultrastructure resembling the membrane attack complex of complement. Proc Natl Acad Sci USA 79: 574–578

Podack ER, Tschopp J, Müller-Eberhard HJ (1982) Molecular organization of C9 within the membrane attack complex of complement. J Exp Med 156:268–282

Podack ER, Young JD, Cohn ZA (1985) Isolation and biochemical and functional characterization of perforin 1 from cytolytic T-cell granules. Proc Natl Acad Sci USA 82:8629–8633

Podack ER, Dahlbäck B, Griffin JH (1986) Interaction of S-protein of complement with thrombin and antithrombin III during coagulation. J Biol Chem 261: 7387–8392

Porter RR, Reid KBM (1978) The biochemistry of complement. Nature 275:699–704

Preissner KT, Wassmuth R, Müller-Berghaus G (1985) Physicochemical characterization of human S-protein and its function in the coagulation system. Biochem J 231:349–355

Prigent D, Alouf JE (1976) Interaction of streptolysin O with sterols. Biochim Biophys Acta 443:288–300

Ramm LE, Mayer MM (1980) Life-span and size of the transmembrane channel formed by large doses of complement. J Immunol 124:2281–2287

Ramm LE, Whitlow MB, Mayer MM (1982a) Transmembrane channel formation by complement: functional analysis of the number of C5b-6, C7, C8, and C9 molecules required for a single channel. Proc Natl Acad Sci USA 79:4751–4755

Ramm LE, Whitlow MB, Mayer MM (1982b) Size of transmembrane channels produced by complement proteins C5b-8. J Immunol 129:1143–1146

Ramm LE, Whitlow MB, Koski CL, Shin ML, Mayer MM (1983) Elimination of complement channels from the plasma membranes of U937, a nucleated mammalian cell line: temperature dependence of the elimination rate. J Immunol 131:1411–1415

Ramm LE, Whitlow MB, Mayer MM (1985) The relationship between channel size and the number of C9 molecules in the C5b-9 complex. J Immunol 134:2594–2599

Roberts WN, Wilson JG, Wong W, Jenkins DE, Fearon DT, Austen KF, Nicholson-Weller An (1985) Normal number and function of CR 1 on affected erythrocytes of patients with paroxysmal noctural hemoglobinuria. J Immunol 134: 512–517

Rogolsky M (1979) Non-enteric toxins of *Staphylococcus aureus*. Microbiol Rev 43:320–360

Rosenfeld SI, Kelly ME, Leddy JP (1976) Hereditary deficiency of the fifth component of complement in man. I. Clinical, immunochemical and family studies. J Clin Invest 57:1626–1634

Rosenfeld SI, Jenkins DE Jr, Leddy JP (1985) Enhanced reactive lysis of paroxysmal nocturnal hemoglobinuria erythrocytes by C5b-9 does not involve increased C7 binding or cell-bound C3b. J Immunol 134:506–511

Rosse WF, Dacie JF (1966) Immune lysis of normal human and paroxysmal nocturnal hemoglobinuria (PNH) red blood cells. I. The sensitivity of PNH red cells to lysis by complement and specific antibody. J Clin Invest 45:736–744

Rosse WF, Adams JP, Thorpe AM (1974) The population of cells in paroxysmal nocturnal haemoglobinuria of intermediate sensitivity to complement lysis: significance and mechanism of increased immune lysis. Br J Haematol 28: 181–190

Rottem S, Cole RM, Habig WH, Barile MF, Hardegree MC (1982) Structural characteristics of tetanolysin and its binding to lipid vesicles. J Bacteriol 152:888–892

Rus HG, Niculescu F, Constantinescu E, Cristea A, Vlaicu R (1986) Immunoelectron-microscopic localization of the terminal C5b-9 complement complex in human atherosclerotic fibrous plaque. Atherosclerosis 61:35–42

Sabatini DD, Kreibich G, Morimoto T, Adesnik M (1982) Mechanisms for the incorporation of proteins in membranes and organelles. J Cell Biol 92:1–22

Sahashi K, Engel AG, Lambert EH, Howard FM (1980) Ultrastructural localization of the terminal and lytic ninth complement component (C9) at the motor end-plate in myasthenia gravis. J Neuropathol Exp Neurol 39:160–172

Salama A, Mueller-Eckhardt C, Boschek B, Bhakdi S (1987) Haemolytic efficiency of C5b-9 complexes in drug-induced immune haemolysis: role of cellular C5b-9 distribution. Br J Haematol (in press, January 1987)

Salant DJ, Belok S, Madaio MP, Couser WG (1980) A new role for complement in experimental membranous nephorpathy in rats. J Clin Invest 66:1339–1350

Sanderson CJ (1976) The mechanism of T cell mediated cytotoxicity. I. The release of different cell components. Proc R Soc Lond [Biol] 192:221–229

Sanderson CJ, Thomas JA (1977) The mechanism of K-cell (antibody-dependent) cell mediated cytotoxicity. I. The release of different cell components. Proc R Soc London [Biol] 197:407–418

Sandvig K, Olsnes S (1981) Rapid entry of nicked diphtheria toxin into cells at low pH. J Biol Chem 256:9068–9076

Sassi F, Hugo F, Muhly M, Kahled A, Ben Rachid MS, Bhakdi S (1987) A natural auto-inhibitory factor of the terminal complement pathway in serum of *Ctenodactylus gondii*. Mol Immunol in press

Schäfer H, Mathey D, Hugo F, Bhakdi S (1986) Deposition of the terminal C5b-9 complement complex in infarcted areas of human myocardium. J Immunol 137:1945–1949

Scharmann W (1976a) Purification and characterization of leucocidin from *Pseudomonas aeruginosa*. J Gen Microbiol 93:292–302

Scharmann W (1976b) Cytotoxic effects of leukocidin from *Pseudomonas aeruginosa* on polymorphonuclear leukocytes from cattle. Infect Immun 13:836–843

Scharmann W, Jacob F, Portstendorfer J (1976) The cytotoxic action of leucocidin from *Pseudomonas aeruginosa* on human polymorphonuclear leukocytes. J Gen Microbiol 93:303–308

Schein SJ, Kagan BL, Finkelstein A (1978) Colicin K acts by forming voltage-dependent channels in phospholipid bilayer membranes. Nature 276:159—161

Schönermark S, Rauterberg EW, Shin ML, Löke S, Roelcke D, Hänsch GM (1986) Homologous species restriction in lysis of human erythrocytes: a membrane-derived protein with C8-binding capacity functions as an inhibitor. J Immunol 136:1772—1776

Sears DA, Weed R, Swisher SN (1964) Differences in the mechanism of in vitro immune hemolysis related to antibody specificity. J Clin Invest 43:975—985

Seeger W, Bauer M, Bhakdi S (1984) Staphylococcal α-toxin elicits hypertension in isolated rabbit lungs due to stimulation of the arachidonic acid cascade. J Clin Invest 74:849—858

Seeger W, Suttorp N, Hellwig A, Bhakdi S (1986) Non-cytolytic terminal complement complexes may serve as calcium gates to elicit leukotriene B4 generation in human polymorphonuclear leukocytes. J Immunol 137:1286—1291

Shany S, Grushoff PS, Bernheimer AW (1973) Physical separation of streptococcal nicotinamide adenine dinucleotide glycohydrolase from streptolysin O. Infect Immun 7:731—734

Shin HS, Pickering RJ, Mayer MM (1971) The fifth component of the guinea-pig complement system. III Dissociation and transfer of C5b, and the probable site of C5b fixation. J Immunol 106:480—493

Shin ML, Paznekas WA, Abramovitz AS, Mayer MM (1977) On the mechanism of membrane damage by complement: exposure of hydrophobic sites on activated complement proteins. J Immunol 119:1358—1364

Shin ML, Hänsch G, Hu VW, Nicholson-Weller A (1986) Membrane factors responsible for homologous species restriction of complement-mediated lysis: evidence for a factor other than DAF operating at the stage of C8 and C9. J Immunol 136:1777—1782

Silversmith RE, Nelsestuen GL (1986) Assembly of the membrane attack complex of complement on small unilamellar phospholipid vesicles. Biochemistry 25:852—860

Simone CB, Henkart PA (1980) Permeability changes induced in erythrocyte ghost targets by antibody-dependent cytotoxic effector cells: evidence for membrane pores. J Immunol 124:954—963

Simone CB, Henkart P (1982) Inhibition of marker influx into complement-treated resealed erythrocyte ghosts by anti-C5. J Immunol 128:1168—1175

Sims PJ (1983) Complement pores in erythrocyte membranes. Analysis of C8/C9 binding required for functional membrane damage. Biochim Biophys Acta 732:541—552

Sims PJ, Lauf PK (1978) Steady state analysis of tracer exchange across the C5b-9 complement lesion in a biological membrane. Proc Natl Acad Sci USA 75:5669—5673

Sims PJ, Lauf PK (1980) Analysis of solute diffusion across the C5b-9 membrane lesion of complement: evidence that individual C5b-9 complexes do not function as discrete, uniform pores. J Immunol 125:2617—2625

Sims PJ, Wiedmer T (1984) Kinetics of polymerization of a flouresceinated derivative of complement protein C9 by the membrane-bound complex of complement proteins C5b-8. Biochemistry 23:3260—3267

Singer SJ, Nicholson GL (1972) The fluid mosaic model structure of cell membranes. Science 175:720—725

Skehel JJ, Baylex PM, Brown EB, Martin SR, Waterfield MD, White JM, Wilson IA, Wiley DC (1982) Changes in the conformation of influenze virus hemagglutinin at the pH-optimum of virus-mediated membrane fusion. Proc Natl Acad Sci USA 79:968—972

Smyth CJ, Duncan JL (1978) Thiol-activated (oxygen-labile) cytolysins. In: Jeljaszewicz J, Waldström T (eds) Bacterial toxins and cell membranes. Academic Press, New York, pp 129—183

Smyth CJ, Freer JH, Arbuthnott JP (1975) Interaction of *clostridium perfringens* theta-haemolysin, a contaminant of commercial phospholipase C with erythrocyte ghost membranes and lipid despersions. A morphological study. Biochim Biophys Acta 382: 479–493

Snyderman RD, Durack T, McCarty GA, Ward RE, Meadows L (1979) Deficiency of the fifth component of complement in human subjects. Am J Med 67:638–645

Springer W, Goebel W (1980) Synthesis and secretion of hemolysin by *Escherichia coli*. J Bacteriol 144:53–59

Stanley KK, Kocher HP, Luzio JP, Jackson P, Tschopp J (1985) The sequence and topology of human complement component C9. EMBO J 4:375–382

Stanley KK, Page M, Campbell AK, Luzio JP (1986) A mechanism for the insertion of complement component C9 into target membranes. Mol Immunol 23:451–458

Steck TL (1974) The organisation of proteins in the human red blood cell membrane. J Cell Biol 62:1–19

Steckel EW, York RG, Monahan JB, Sodetz JM (1980) The eighth component of human complement. J Biol Chem 255:11997–12005

Steckel EW, Welbaum BE, Sodetz JM (1983) Evidence of direct insertion of terminal complement proteins into cell membrane bilayers during cytolysis. J Biol Chem 258:4318–4324

Stewart JL, Monahan JB, Brickner A, Sodetz JM (1984) Measurement of the ratio of the eighth and ninth components of human complement on complement-lysed membranes. Biochemistry 23:4016–4022

Stolfi RL (1968) Immune lytic transformation: a state of irreversible damage generated as a result of the reaction of the eighth component in the guinea pig complement system. J Immun 100:46–54

Suttorp N, Seeger W, Dewein E, Bhakdi S, Roka L (1985a) Staphylococcal α-toxin stimulates synthesis of prostacyclin by cultured endothelial cells from pig pulmonary arteries. Am J Physiol 248:C127–135

Suttorp N, Seeger W, Uhl J, Lutz F, Roka L (1985b) *Pseudomonas aeruginosa* cytotoxin stimulates prostacyclin production in cultured pulmonary artery endothelial cells: membrane attack and calcium influx. J Cell Phys 123:64–72

Tanford C, Reynold JA (1976) Characterization of membrane proteins in detergent solution. Biochim Biophys Acta 457:133–169

Tedesco F, Bardare M, Giovanetti AM, Sirchia G (1980) A familial dysfunction of the eighth component of complement (C8). Clin Immunol Immunopathol 16:180–191

Tedesco F, Densen P, Villa MA, Peterson BH, Sirchia G (1983) Two types of dysfunctional eighth component of complement molecules in C8 deficiency in man: reconstitution of normal C8 from the mixture of two abnormal C8 molecules. J Clin Invest 71:183–191

Thelestam M, Möllby R (1975) Sensitive assay for detection of toxin-induced damage to the cytoplasmic membrane of human diploid fibroblasts. Infect Immun 12:225–232

Thelestam M, Möllby R, Wadström T (1973) Effects of staphylococcal alpha-, beta-, delta-, gamma-hemolysins on human diploid fibroblasts and HeLa cells. Evaluation of a new quantitative assay for measuring cell damage. Infect Immun 8:938–946

Thelestam M, Jolivet-Reynaud C, Alouf JE (1983) Photolabeling of staphylococcal α-toxin from within rabbit erythrocyte membranes. Biochem Biophys Res Comm 111:444–449

Thompson RA, Lachmann PJ (1970) Reactive lysis: the complement-mediated lysis of unsensitized cells. I. The characterization of the indicator factors and its identification as C7. J Exp Med 131:629–643

Thompson RA, Rowe DS (1968) Reactive haemolysis – a distinctive form of red cell lysis. Immunology 14:745–750

Tobkes N, Wallace BA, Bayley H (1985) Secondary structure and assembly mechanism of an oligomeric channel protein. Biochemistry 24:1915–1920

Todd EW (1938b) Lethal toxins of hemolytic streptococci and their antibodies. Br J Exp Pathol 19:367–378

Tosteson DC, Andreoli TE, Tiefenberg M, Cook P (1968) The effects of macrocyclin compounds on cation transport in sheep red cells and thin and thick lipid membranes. J Gen Physiol 51:373s–384s

Tranum-Jensen J (1987a) Electron microscopical assays: negative staining. Methods Enzymol (in press)

Tranum-Jensen J (1987b) Electron microscopical assays: freeze-fracture and -etching. Methods Enzymol (in press)

Tranum-Jensen J, Bhakdi S (1983) Freeze-fracture ultrastructural analysis of the complement lesion. J Cell Biol 97:618–626

Tranum-Jensen J, Bhakdi S, Bhakdi-Lehnen B, Bjerrum OJ, Speth V (1978) Complement lysis: the ultrastructure and orientation of the C5b-9 complex on target sheep erythrocyte membranes. Scand J Immun 7:45–56

Tschopp J (1984) Circular polymerization of the membranolytic ninth component of complement: dependence on metal ions. J Biol Chem 259:10569–10573

Tschopp J, Esser AF, Spira TJ, Müller-Eberhard HJ (1981) Occurrence of an incomplemente C8 molecule in homozygous C8 deficiency in man. J Exp Med 154:1599–1607

Tschopp J, Müller-Eberhard HJ, Podack ER (1982a) Formation of transmembrane tubules by spontaneous polymerization of the hydrophilic complement protein C9. Nature 298:534–538

Tschopp J, Müller-Eberhard HJ, Podack ER (1982b) Ultrastructure of the membrane attack complex of complement: detection of the tetramolecular C9-polymerizing complex C5b-8. Proc Natl Acad Sci USA 79:7474–7478

Tschopp J, Engel A, Podack ER (1984) Molecular weight of poly C9 of human complement. J Biol Chem 259:1922–1928

Tschopp J, Podack ER, Müller-Eberhard HJ (1985) The membrane attack complex of complement: C5b-8 complex as accelerator of C9 polymerization. J Immunol 134:495–499

Unwin N (1986) Is there a common design for cell membrane channels? Nature 323:12–13

Urry DW (1972) Protein conformation in biomembranes: optical rotation and absorption of membrane suspensions. Biochim Biophys Acta 265:115–168

Valet G, Opferkuch W (1975) Mechanism of complement-induced cell lysis. Demonstration of a three-step mechanism of EAC1-8 lysis by C9 and of a non-osmotic swelling of erythrocytes. J Immunol 115:1028–1033

van den Bosch JF, Emody L, Ketyi I (1982a) Virulence of haemolytic strains of Escherichia coli in various animal models. FEMS Microbiol Lett 13:427–430

van den Bosch JF, Postma P, Koopman PAR, de Graaff J, MacLaren CM, van Brenk DG, Guinee PAM (1982b) Virulence of urinary and faecal Escherichia coli in relation to serotype, haemolysis and haemagglutination. J Hyg 88:567–577

van Epps D, Andersen BR (1969) Streptolysin O: sedimentation coefficient and molecular weight determinations. J Bacteriol 100:526–527

van Epps D, Andersen BR (1973) Isolation of streptolysin O by preparative polyacrylamide gel electrophoresis. Infect Immun 7:493–495

van Epps D, Andersen BR (1974) Streptolysin O inhibition of neutrophil chemotaxis and mobility: non immune phenomenon with species specificity. Infect Immun 9:27–33

Vlaicu R, Niculescu F, Rus HG, Cirstea A (1985) Immunohistochemical localization of the terminal C5b-9 complement complex in human aortic fibrous plaque. Atherosclerosis 57:163–177

Vogt W, Schmidt G, von Buttlar B, Dieminger L (1978) A new function of the activated third component of complement: binding to C5, an essential step for C5 activation. Immunology 34:29–40

Waalwijk C, MacLaren DM, de Graaff J (1983) In vivo function of hemolysin in the nephropathogenicity of *Escherichia coli*. Infect Immun 42:245—249

Wagner W, Vogel M, Goebel W (1983) Transport of hemolysin across the outer membrane of *Escherichia coli* requires two functions. J Bacteriol 154:200—210

Ward RHR, Lachmann PJ (1985) Monoclonal antibodies which react with lymphocyte-lysed target cells and which crossreact with complement-lysed ghosts. Immunology 56:179—188

Ware CF, Kolb WP (1981) Assembly of the functional membrane attack complex of human complement: formation of disulfide-linked C9 dimers. Proc Natl Acad Sci USA 78:6426—6430

Ware CF, Wetsel RA, Kolb WP (1981) Physicochemical characterization of fluid phase (SC5b-9) and membrane derived (MC5b-9) attack complexes of human complement purified by immunoadsorbent affinity chromatography or selective detergent extraction. Mol Immunol 18:521—531

Watson KC, Kerr EJC (1974) Sterol structural requirements for inhibition of streptolysin O activity. Biochem J 140:95—98

Weiner RN, Reinacher M (1982) Lower nephron toxicity of a highly purified cytotoxin from *Pseudomonas aeruginosa* in rats. Exp Mol Pathol 37:249—271

Welch RA, Falkow S (1984) Characterization of *Escherichia coli* hemolysins conferring quantitative differences in virulence. Infect Immun 43:156—160

Welch RA, Dellinger EP, Minshew B, Falkow S, (1981) Haemolysin contributes to virulence of extra-intestinal *E. coli* infections. Nature 294:665—667

Whitlow MB, Ramm LE, Mayer MM (1985) Penetration of C8 and C9 in the C5b-9 complex across the erythrocyte membrane into the cytoplasmic space, J Biol Chem 260:998—100

Wilkinson PC (1975) Inhibition of leukocyte locomotion and chemotaxis by lipid-specific bacterial toxins. Nature 255:485—487

Yamamoto K, Kawshima T, Migita S (1982) Glutathione-catalyzed disulfide linking C9 in the membrane attack complex of complement. J Biol Chem 257, 8573—8576

Young JDE, Young TM, Lu LP, Unkeless JC, Cohn ZA (1982) Characterization of a membrane pore-forming protein from *Entamoeba histolytica*. J Exp Med 156:1677—1690

Young JDE, Hengartner H, Podack ER, Cohn ZA, (1986a) Purification and characterization of a cytolytic pore-forming protein from granules of cloned lymphocytes with natural killer activity. Cell 44:849—859

Young JDE, Nathan CF, Podack ER, Palladino MA, Cohn ZA (1986b) Functional channel formation associated with cytotoxic T cell granules. Proc Natl Acad Sci USA 83:150—154

Young JDE, Podack ER, Cohn ZA (1986c) Properties of a purified pore-forming protein (perforin 1) isolated from H2-restricted cytotoxic T-cell granules. J Exp Med 164:144—155

Young JDE, Leong LG, Liu C, Damiano A, Cohn ZA (1986d) Extracellular release of lymphocyte cytolytic pore-forming protein (perforin) after ionophore stimulation. Proc Natl Acad Sci USA 83:5668—5672

Zalman LS, Brothers MA, Chiu FJ, Müller-Eberhard HJ (1986) Mechanism of cytotoxicity of human large granular lymphocytes: relationship of the cytotoxic lymphocyte protein to the ninth component (C9) of human complement. Proc Natl Acad Sci USA 83:5262—5266

Subject Index